AF540279

OPPORTUNITY IN CHEMISTRY

OPPORTUNITY IN CHEMISTRY

By
Dr. Shardendu Kislaya

DISCOVERY PUBLISHING HOUSE PVT. LTD.
NEW DELHI-110 002

Published by:
Tilak Wasan
DISCOVERY PUBLISHING HOUSE PVT. LTD.
4831/24, Ansari Road, Prahlad Street
Darya Ganj, New Delhi-110002 (India)
Phone: +91-11-23279245, 43764432
Fax: +91-11-23253475
E-mail: parul.wasan@gmail.com
info@discoverypublishinggroup.com
discoverypublishinghouse@gmail.com
web: www.discoverypublishinggroup.com

***First Edition:* 2011**
ISBN: 978-81-8356-806-7

Opportunity in Chemistry

Printed at:
Mehra Offset Press
Delhi

Preface

The book "Opportunity in Chemistry " has been written in general properties of the subject matter. This change is that paper division in the examination is different Universities. One of salient course syllabi features of the present edition of the book is the inclusion of a large number of typical worked out problems which will elucidate various abstract principles and theories discussed in the text. It covers the complete syllabus of physics prescribed by Technical Universities. The treatment given is simple, lucid and comprehensive. The language used is simple.

The classification of the questions, specially the short-answer type questions, will, I hope, help the students to prepare for the final examination. I have spared no pains in applying my long experience of degree class teaching to make the book useful to the students.

I shall, therefore, consider, my labours amply rewarded if the book receives warm reception and appreciation from the teachers and the taught. Though nothing can be claimed as original but the subject matter has been arranged in my own style.

— *Author*

Contents

Pages

Preface

1. **Opportunity in Chemistry** 1

An Introduction, Characteristics of Matter, Characteristics of Gases, Laws of the Gas, Gay Lussac's Law, Graham's Law of Diffusion, Molecular Velocity, Kinds of Velocities, Collision Properties of Molecules, Determination of Critical Constants, Liquefaction of Gases, Andrews Isotherms of Carbon Dioxide, Methods of Liquefaction of Gases, Electric Nature of Molecules, Molar Polarization, Molecular Dipole Moments, Magnetic Properties of Molecules, Diamagnetism and Paramagnetism, Molecular Energies, Energies of the Diatomic Molecules of a Gas, Classical Vibrations, Angular Momentum, Rotational Energies, Three-Dimensional Translational Energy.

2. **Radioactivity** 35

Introduction, Use of the Radioactivity, Nuclear Energy, Important Method of Radioactivity, Properties of Radiations, Nuclear Reactions, Nuclear Fission Reactions, Nuclear Fusion Reactions, Nuclear Reactors, Mole Fraction, Henry's Law, Solutions of Liquids, Steam Distillation, Dissolution and Recrystallisation, Energies of Atoms and Molecules, Partition Function, Rotational and Vibrational Motions, Heat Capacities, Hardness of Crystals, Hardness of Crystals, Measurement of Vapour Pressure, Boiling Point Elevation, Measurement of

Boiling Point Elevation, Beckmann's Method, Osmotic Pressure, Gibbs Duhem Equation, Partial Molal Quantities, Quantum Mechanics and Wave Mechanics, The Uncertainty Principle, Cathode Rays, Bohr Atomic Model, Atomic Spectra, Photoelectric Effect, Zeeman Effect, Self Consistent Field Method, Electron Configuration, Hydrogen and Helium, Electronegativity, Orientation Quantum Number, Molecular Treatment, Osmosis and Diffusion, Osmotic Pressure, Osmotic Pressure Determinations of Molecular Masses, Osmosis Theories, Membrane Bombardment Theory, Van't Hoff Theory, Sedimentation Coefficient, Sedimentation Velocity, Electrophoresis, Isoionic Points.

3. Free Energy 106

Introduction, Free Energy Function, Free Energy of Real Gases, Interrelation of Thermodynamic Properties, Free Energy and Temperature, Concept of Hydrolysis, Hydrolysis Constant, Determination of Degree of Hydrolysis, Electrochemical Cells, EMF and Activities, Electrolyte-concentration, Ion-Selective Membrane, Ostwald's Law, Degree of Dissociation, Solubility Equilibria and the Solubility Product, Selective Precipitation, Colloidal Systems, Typical Sols, Preparation of Lyophobic Sols, Kinetic Properties of Sols, Applications of Colloids, Emulsions.

4. Specific Conductance 146

Introduction, Types of Electrolytes, Transference Numbers, Calculation of Ionic Mobilities, Ionic Conductance, Ions of a Weak Electrolyte, Chemical Bonding, Ionic Bond, Covalent Bond, Hydrogen Bonding, Molecular Orbitals, Metallic Bonding, Electron Negativity, Virial Theorem, The Pyramidal and Angular Molecules, Mechanisms of Chemical Reactions, The Rate of Reaction, Order of a Reaction, Molecularity of a Reaction, First Order Reactions, Second Order Reaction, Third Order Reactions, Determination of The Order of a

Reaction, Complex Reactions, Reversible or Opposing Reactions, Catalysis, Catalytic Reactions, Enzyme Catalysis, Theories of Catalysis, Photochemical Reactions, Primary and Secondary Reactions, Photosensitized Reaction.

5. **Types of Solids in Chemistry** 200

Introduction, Structure of a Crystal, Bragg's Equation, Miller Indices, The Diffraction Spots, Single Crystal Diffraction and Powder Diffraction, Arrangement of Ions in the Unit Cell, Classification of Crystals, The Common Crystal Defects, Neutron Diffraction, Ultramicroscope, The Liquid Crystals, First Law of Thermodynamics, The Energy of a Chemical System, Non-Equilibrium States, Isothermal Expansion of a Gas, First Law of Thermodynamics, Heat Capacities, Thermal Part of the Internal Energy, Ionic Hydration, Enthalpy Change in Chemical Reactions, Joule Thomson Experiment, Heat and Work, Second Law of Thermodynamics, Entropy, The Carnot Cycle, Some Spontaneous Processes, Third Law of Thermodynamics, Fugacity and Activity, Work and Free Energy Functions.

6. **Raman Spectroscopy** 242

Introduction, Spectra of Conjugated Systems, Nuclear Magnetic-Resonance Spectroscopy, Electron Spin Spectroscopy, Phase Rule, Pressure Temperature Phase Diagrams, Phase Equilibria, Determination of Transition Point, Three Component Systems, Zinc-Cadmium System, The Magnesium-Zinc System, Adsorption, Adsorption Isotherm, Langmuir Adsorption Isotherm, Adsorption Isotherm Liquids, Applications of Adsorptions.

1

Opportunity in Chemistry

AN INTRODUCTION

Opportunity and Needs, edited by Frank H. Westheimer. The Frontiers that it envisaged in 1965 have indeed fulfilled their promise, and almost all of the report's optimistic expectations have been realized.

In fact, the advances have been so rapid and so penetrating that the structure of chemistry and its interactions with contiguous disciplines have qualitatively changed. Physics has provided a panoply of powerful diagnostic tools that extend the experimental horizons of chemistry.

Any attempt to learn about the internal structure and properties of individual molecules might seem to be too bold an undertaking. We begin by setting aside worries about the small size of molecules. Molecules will be treated as ordinary, but little, things. They do, however, exhibit behavior that is quite outside our ordinary experience. We must treat these molecular-world particles by the methods of *quantum*, or *wave mechanics*.

Here you are introduced to the methods that are used and the results that are obtained when quantum mechanics is applied to the study of gas molecules. This will lead you to a new way of thinking about these molecules. You will take an important step into the molecular world as you extend and refine the billiard-ball view of molecules that the simple kinetic-molecular theory encourages.

CHARACTERISTICS OF MATTER

All matter exists in three states; gas, liquid and solid, (the fourth one is plasma.)

A *gas* consists of molecules separated wide apart in empty space. The molecules are free to move about throughout the container.

A *liquid* has molecules touching each other. However, the intermolecular space, permit the movement of molecules throughout the liquid.

A *solid* has molecules, atoms or ions arranged in a certain order in fixed positions in the crystal lattice. The particles in a solid are not free to move about but vibrate in their fixed positions.

Of the three states of matter, the gaseous state is the one most studied and best understood. We shall consider it first.

CHARACTERISTICS OF GASES

Expansibility : Gases have limitless expansibility. They expand to fill the entire vessel they are placed in.

Compressibility : Gases are easily compressed by application of pressure to a movable piston fitted in the container.

Diffusibility : Gases can diffuse rapidly through each other to form a homogeneous mixture.

Pressure : Gases exert pressure on the walls of the container in all directions.

When a gas, confined in a vessel is heated, its pressure increases. Upon heating in a vessel fitted with a piston, volume of the gas increases.

The above properties of gases can be easily explained by the Kinetic Molecular Theory which will be considered later in the chapter.

A gas sample can be described in terms of four parameters.

(1) The volume, V of the gas

(2) Its pressure, P

(3) Its temperature, T

(4) The number of moles, n, of gas in the container.

The volume of the container is the volume of the gas sample. It is usually given in litre (L) or millilitres (mL).

$$1l = 1000ml \text{ and } 1ml = 10^{-3}l$$

One millilitre is practically equal to one cubic centimetre (cc). Actually,

$$1l = 1000.028cc$$

The S.I. unit for volume is cubic metre (m^3) and the smaller unit is decimeter^3 (dm^3).

The pressure of a gas is defined as the force exerted by the impacts of its molecules per unit surface area in contact

The pressure of a gas sample can be measured with the help of a mercury manometer. Similarly, the atmospheric pressure can be determined with a mercury barometer .

The pressure of air, that can support 760 mm Hg. Column at sea level is called one atmosphere. The unit of pressure millimetre of mercury is called torr.

The S.I. unit of pressure is the Pascal (Pa). The relation between atmosphere, torr and Pascal is

$$2 \text{ atm} = 760 \text{ torr} = 1.013 \times 10^5 \text{ Pa}.$$

But the unit of pressure Pascal is not in common use.

The temperature of a gas may be measured in centigrade degrees or celsius degrees. The S.I. unit of temperature is Kelvin (k). By using equation,

$$k = C^o + 273,$$

the centigrade degrees can be converted to Kelvin. The moles of gas can be calculated as follows :

$$\text{Moles of Gas}(n) = \frac{\text{Mass of gas sample}(m)}{\text{Molecular mass of gas }(M)}$$

LAWS OF THE GAS

During 18th century scientists derived the relationships, among the pressure temperature, and volume of a given mass of a gas. These relationships which describe the general behavior of gases are called the gas laws.

In 1660 Robert Boyle found out experimentally the change in the volume of a given sample of gas with pressure at room temperature. He formulated a generalisation known *Boyle's Law*. His law states that.

At constant temperature, the volume of a fixed mass of gas is inversely proportional to its pressure.

$$V = k \times \frac{1}{P} \text{ (k = Proportionally constant).}$$

In 1787 Jacques Charle's investigated the effect of change of temperature on the volume of a fixed amount of gas at constant pressure. Later his generalisation is called, the Charle's law. It states that.

At constant pressure the volume of a fixed mass of gas is directly proportional to the Kelvin temperature or absolute temperature. Mathematically the law can be expressed as follows:

$$V \propto T \text{ (P, n are constant)}$$

$$V = kT \text{ (k = constant)}$$

Or $\frac{V}{T} = k.$

If V_1, T_1 are the initial volume and temperature of a given mass of gas at constant. Pressure and V_2, T_2 be the new values then it can be written as follows:

$$\frac{V_1}{T_1} = k = \frac{V_2}{T_2}$$

or $$\frac{V_1}{T_1} = \frac{V_2}{T_2}.$$

Bolye's Law and Charle's Law can be combined into a single relationship called combined gas law.

Boyle's law $V \propto \frac{1}{P}$ (T, n constant)

Charles law $V \propto T$ (P, n are constant)

Therefore $V \propto \frac{T}{P}$ (n constant).

The combined law can be stated as:

for a fixed mass of gas, the volume is directly proportional to kelvin temperature and inversely proportional to the Pressure.

If k be the proportionality constant,

$$V = \frac{kT}{P} \quad \text{(n constant)}$$

or $$\frac{PV}{T} = k \quad \text{(n constant)}$$

If the pressure, volume and temperature of a gas be changed from P_1, V_1, and T_1, to P_2 T_2 and V_2, then

$$\frac{P_1V_1}{T_1} = k,$$

$$\frac{P_2V_2}{T_2} = k$$

$$\frac{P_1V_1}{T_1} = \frac{P_2V_2}{T_2}$$

This is the form of combined law for two sets of conditions. It can be used to solve problems involving a change in the three variables P, V and T for a fixed mass of gas.

Example:

25.8 litre of a gas has a pressure of 690 torr and a temperature of 17°C. What will be the volume if the pressure is changed to 1.85 aim and the temperature to 345 K.

Solution:

Initial Conditions :	*Final Conditions :*
$V_1 = 25.8$ litres	$V_2 = ?$
$P_1 = \frac{690}{7600} = 0.908$ atm	$P_2 = 1.85$ atm
$T_1 = 17 + 273 = 290$K	$T_2 = 345$K.

Substituting values in the equation

$$\frac{P_1V_1}{T_1} = \frac{P_2V_2}{T_2}$$

$$\frac{0.908 \times 25.8 \text{ liter}}{290\text{K}} = \frac{1.85\text{atm} \times V_2}{345\text{K}}$$

Whence $$V_2 = \frac{0.908 \times 25.8}{290 \times 1.85} = 15.1 \text{ litres.}$$

GAY LUSSAC'S LAW

In 1802 Joseph Gay Lussac as a result of his experiments established a general relation between the pressure and temperature of a gas. This is known as *Gay Lussac's Law* or *Pressure-Temperature Law*. It states that:

at constant volume, the pressure of a fixed mass of gas is directly proportional to the kelvin temperature or absolute temperature.

The law may be expressed mathematically as

$$P \propto T \qquad \text{(Volume, n are constant)}$$

or $$P = kT$$

or $$\frac{P}{T} = k$$

For different conditions of pressure and temperature

$$\frac{P_1}{T_1} = k = \frac{P_2}{T_2}$$

or $$\frac{P_1}{T_1} = \frac{P_2}{T_2}$$

Knowing P_1, T_1, and T_2, P_2 can be calculated.

Avogadro's Law

Let us take a balloon containing a certain mass of gas. If we add to it more mass of gas, holding the temperature (T) and pressure (P) constant, the volume of gas (V) will increase. It was found experimentally that the amount of gas in moles is proportional to the volume. That is,

$$V \propto n \qquad \text{(T and P constant)}$$

or $$V = A\,n$$

where A is constant of proportionality.

or $$\frac{V}{n} = A$$

For any two gases with volumes V_1, V_2 and moles n_1, n_2 at constant T and P,

$$\frac{V_1}{n_1} = A = \frac{V_2}{n_2}$$

If $$V_1 = V_2, \quad n_1 = n_2.$$

Thus, for equal volumes of the two gases at fixed T and P, number of moles is also equal. This is the basis of *Avogadro's Law* which may be stated as :*equal volumes of gases at the same temperature and pressure contain equal number of moles or molecules.*

The Molar Gas Volume

It follows as a corollary of Avogadro's Law that one mole of any gas at a given temperature (T) and pressure (P) has the same fixed volume. It is called the *molar gas volume* or *molar volume*. In order to compare the molar volumes of gases, chemists use a fixed reference temperature and pressure. This is called standard temperature and pressure (abbreviated, *STP*). The standard temperature used is 273 K (0°C) and the standard pressure is 1 atm (760 mm Hg). At STP we find experimentally that one mole of any gas occupies a volume of 22.4 litres.

If we take it in the form of equation, we get,

1 mole of a gas at STP = 22.4 litres.

GRAHAM'S LAW OF DIFFUSION

When two gases are placed in contact, they mix spontaneously. This is due to the movement of molecules of one gas into the other gas. This process of mixing of gases by random motion of the molecules is called *Diffusion*. Thomas Graham observed that molecules with smaller masses diffused faster than heavy molecules.

In 1829 Graham formulated what is now known as Graham's Law of Diffusion. It states that: *under the same conditions of temperature and pressure, the rates of diffusion of different gases are inversely proportional to the square roots of their molecular masses.*

Mathematically the law can be expressed as

$$\frac{r_1}{r_2} = \sqrt{\frac{M_2}{M_1}}$$

where r_1 and r_2 are the rates of diffusion of gases 1 and 2, while M_1 and M_2, are their molecular masses.

When a gas escapes through a pin-hole into a region of low pressure or vacuum, the process is called *Effusion*. The rate of effusion of a gas also depends, on the molecular mass of the gas. Dalton's law can also be applied to effusion of a gas.

Dalton's law when applied to effusion of a gas is called die Dalton's Law of Effusion. It may be expressed mathematically as

$$\frac{\text{Effusion rate of Gas 1}}{\text{Effusion rate of Gas 2}} = \sqrt{\frac{M_2}{M_1}} \qquad \text{(P, T constant)}$$

The determination of rate of effusion is much easier compared to the rate of diffusion. Therefore, Dalton's law of effusion is often used to find the molecular mass of a given gas.

Example (a):

If a gas diffuses at a rate of one-half as fast as O_2 find the molecular mass of the gas.

Solution:

Applying Graham's Law of Diffusion,

$$\frac{r_1}{r_2} = \sqrt{\frac{M_2}{M_1}}$$

$$\frac{1/2}{1} = \sqrt{\frac{32}{M_1}}$$

Squaring both sides of the equation.

$$\left(\frac{1}{2}\right)^2 = \frac{32}{M_1}$$

or $$\frac{1}{4} = \frac{32}{M_1}$$

Whence, $M_1 = 128.$

Thus, the molecular mass of the unknown gas is **128**.

Example (b):

50 ml of gas A effuse through a pin-hole in 146 seconds. The same volume of CO_2 under identical conditions effuses in 115 seconds. Calculate the molecular mass of A.

Solution:

$$\frac{\text{Effusion rate of } CO_2}{\text{Effusion rate of A}} = \sqrt{\frac{M_A}{M_{CO_2}}}$$

$$\frac{50/115}{50/146} = \sqrt{\frac{M_A}{44}}$$

or $$(1.27)^2 = \frac{M_A}{44}$$

Whence $M_A = 71.$

∴ Molecular mass of A is **71**.

MOLECULAR VELOCITY

When any two molecules collide, one molecule transfers kinetic energy ($1/2mv^2$) to the other molecule. The velocity of the molecule which gains energy increases and that of the other decreases. Millions of such molecular collisions are taking place per second. Therefore, the velocities of molecules are changing constantly. Since the number of molecules is very large, a fraction of molecules will have the same particular velocity. In this way there is a broad distribution of velocities over different fractions of molecules. In 1860 James Clark Maxwell calculated the distribution of velocities from the laws of probability. He derived the following equation for the distribution of molecular velocities.

$$\frac{dN_c}{N} = 4\pi\left(\frac{M}{2\pi RT}\right)^{3/2} e^{\frac{-MC^2}{2RT}} C^2 dc$$

where

dN_c = number of molecules having velocities between C and (C+dc)

N = total number of molecules

M = molecular mass

T = temperature on absolute scale (K).

The relation stated above is called *Maxwell's law of distribution of velocities*. The ratio dn_c/n gives the fraction of the total number of molecules having velocities between C and (C + dc). Maxwell plotted such fractions against velocity possessed by the molecules. The curves so obtained illustrate the salient features of Maxwell distribution of velocities.

The distribution of velocities in nitrogen gas, N_2, at 300 K and 600K. It will be noticed that:

1. A very small fraction of molecules has either very low (close to zero) or very high velocities.
2. Most intermediate fractions of molecules have velocities close to an average velocity represented by the peak of the curve. This velocity is called the *most probable velocity*. It may be defined as the velocity possessed by the largest fraction of molecules corresponding to the highest point on the Maxvellian curve.

At higher temperature, the whole curve shifts to the right (dotted curve at 600 K). This shows that *at higher temperature more molecules have higher velocities and fewer molecules have lower velocities,*

KINDS OF VELOCITIES

Different kinds of molecular velocities can be studied as follows:

1. The Average Velocity (v)
2. The Root Mean Square Velocity (μ)
3. The Most Probable Velocity (v_).

1. The Average Velocity

Let mere be n molecules of a gas having individual velocities v_1, v_2, v_3 .... v_n. The ordinary average velocity is the arithmetic mean of the various velocities of the molecules.

$$\bar{v} = \frac{v_1 + v_2 + v_3 .... + v_n}{n}$$

From Maxwell equation it has been established that the average velocity v_{av} is given by the expression

$$\bar{v} = \sqrt{\frac{8RT}{\pi M}}$$

Substituting the values of R, T, π and M in this expression, the average value can be calculated.

2. Root Mean Square Velocity

If v_1, v_2, v_3 .... v_n are the velocities of n molecules in a gas, v^2, the mean of the squares of all the velocities is

$$\mu^2 = \frac{v_1^2 + v_2^2 + v_3^2 .... + v_n^2}{n}$$

Taking the root $$\mu = \sqrt{\frac{v_1^2 + v_2^2 + v_3^2 .... + v_n^2}{n}}$$

μ is thus the *Root Mean Square Velocity* or *RMS Velocity*. It is denoted by u.

The value of the RMS of velocity u, at a given temperature can be calculated from the Kinetic energy,

$$PV = \frac{1}{3} mN\mu^2$$

$$\mu^2 = \frac{3PV}{MN}$$

For one mole gas $PV = RT$

$$\mu^2 = \frac{3RT}{M}$$

$$\mu = \sqrt{\frac{3RT}{M}}.$$

By substituting the values R, T and M the value of u can be determined with the help of u, the total kinetic energy of a gas can be calculated.

3. The Most Probable Velocity

Most probable velocity is possessed by the largest number of molecules in a gas. According to the calculations made by Maxwell, the most probable velocity

v_{mp}, is given by the expression

$$v = \sqrt{\frac{2RT}{M}}$$

Substituting the values of R, T, and M in this expression, the most probable velocity can be calculated.

The velocities of gas molecules are exceptionally high. Thus, velocity of hydrogen molecule is 1,838 metres sec^{-1}. While it may appear impossible to measure so high velocities, these can be easily calculated from the Kinetic Gas equation. Several cases may arise according to the available data.

While calculating different types of velocities, we can also make use of the following expressions stated already.

RMS velocity, $\mu = \sqrt{\frac{3RT}{M}}$

Average velocity, $\bar{v} = \sqrt{\frac{8RT}{M}}$

Most Probable velocity, $v_{mp} = \sqrt{\frac{2RT}{M}}.$

COLLISION PROPERTIES OF MOLECULES

The molecules in a gas are constantly colliding with one another. Now let us study about certain Collision Properties of Molecules.

The Mean Free Path

At a given temperature, a molecule travels in a straight line before collision with another molecule. The distance travelled by the molecule before collision is termed free path. The free path for a molecule varies from time to time. *The mean distance travelled by a molecule between two successive collisions is called the Mean Free Path.* It is denoted by λ. If l_1, l_2, l_3 are the free paths for a molecule of a gas, its free path

$$\lambda = \frac{l_1 + l_2 + l_3 + \ldots + l_n}{n}$$

where n is the number of molecules with which the molecule collides. Evidently, the number of molecular collisions will be less at a lower pressure or lower density and longer will be the mean free path. The mean free path is also related with the viscosity of the gas.

The mean free path, λ, is given by the expression

$$\lambda = \eta\sqrt{\frac{3}{Pd}}$$

where P = pressure of the gas

d = density of the gas

η = coefficient of viscosity of the gas.

By a determination of the viscosity of the gas, the mean free path can be readily calculated. At STP, the mean free path for hydrogen is 1.78×10^{-5} cm and for oxygen it is 1.0×10^{-5} cm.

Now let us study about the effect of temperature and pressure on Mean Free Path.

Temperature

The ideal gas equation for, n moles of a gas is

$$PV = nRT \qquad \ldots(i)$$

where 'n' is the number of moles given by

$$n = \frac{\text{Number of molecules}}{\text{Avagadro's number}} = \frac{N}{N_0}$$

When we substitute this in equation (i), we get

$$PV = \frac{N}{N_0}RT$$

$$\frac{N}{V} = \frac{PN_0}{RT}$$

At constant pressure $N \propto \frac{1}{T}$ ...(ii))

The mean free path is given by

$$\lambda = \frac{\text{Distance travelled by the molecule per second}}{\text{Number of collisions per c.c.}}$$

$$= \frac{\bar{v}}{\sqrt{2}\pi\sigma^2\bar{v}\bar{N}}$$

$$= \frac{1}{\sqrt{2}\pi\sigma^2\bar{N}} \quad \text{...(iii)}$$

combining equation (ii), and (iii) we get

$$\lambda \propto T$$

Thus, *the mean free path is directly proportional to the absolute temperature.*

Pressure

We know that the pressure of a gas at certain temperature is directly proportional to the number of molecules per c.c. *i.e.,*

$$P \propto \bar{N}$$

and mean free path is given by

$$\lambda = \frac{1}{\sqrt{2}\lambda\sigma^2\bar{N}}$$

Combining there two equations, we get

$$\lambda \propto \frac{1}{p}$$

Thus, *the mean free path of a gas is directly proportional to the pressure of a gas at constant temperature.*

Example:

At 0°C and 1 atmospheric pressure the molecular diameter of a gas is 4A.. Calculate the mean free path of its molecule.

Solution

The mean free path is given by

$$\alpha = \frac{1}{\sqrt{2}\pi\sigma^2\overline{N}}$$

where σ is the molecular diameter

and $\overline{N}$ is the no. of molecules per c.c.

Here $\sigma = 4\text{Å} = 4 \times 10^{-8}$ cm.

We know 22400 ml of a gas 0°C and 1 atm. pressure contains 6.02×10^{23} molecules.

$$\therefore \text{ No. of molecules per c.c., } \overline{N} = \frac{6.02 \times 10^{23}}{22400}$$

$$= 2.689 \times 10^{19} \text{ molecules.}$$

When we substitute these values we get,

$$\sigma = \frac{1}{1.414 \times 3.14 \times (4 \times 10^{-8})^2 \times 2.689 \times 10^{19}}$$

$$= \frac{1}{1.414 \times 3.14 \times 16 \times 2.689 \times 10^3}$$

$$= 0.524 \times 10^{-5} \text{ cm.}$$

DETERMINATION OF CRITICAL CONSTANTS

The actual determination of critical constants of a substance is often a task of considerable difficulty. Of these the critical temperature and critical pressure can be measured relatively easily with the help of *Cagniard de la Tour's apparatus*. It consists of a stout glass U-tube blown into a bulb at the lower end. The liquid under-examination is contained in the bulb and the rest of the apparatus is filled with mercury. The upper end of the U-tube is sealed leaving a little air in it so that this can be used as a manometer.

The temperature of the bulb containing the liquid and its vapour is raised gradually by means of the heating jacket. A point is reached when the meniscus of the liquid becomes faint and then disappears leaving the contents of the bulb perfectly homogeneous. On allowing the bulb to cool again, a mist first forms in the gas which quickly settles with the reappearance of the meniscus. *The mean of the temperatures of disappearance and reappearance of the meniscus in the bulb, is the critical temperature*. The pressure read on the manometer at the critical temperature, gives the critical pressure.

The critical volume is the volume at critical temperature and critical pressure. It is much more difficult to measure since even a slight change in temperature of pressure at the critical point produces a large change in volume. When the critical volume is determined, critical density also can be easily found out. Now, let us study about the illustration of the determination of critical volume.

The mean values of the densities can be plotted against the various temperature, where straight line DC is obtained which will obviously pass through the critical temperature, which will be given by the point, where this line will cut the curve AB. The density corresponding to the point C in the diagram is the critical density.

The following table gives the critical temperature and critical pressure of a few substances.

LIQUEFACTION OF GASES

A gas can be liquefied by lowering the temperature and increasing the pressure. At lower temperature, the gas molecules lose kinetic energy. The slow moving molecules then aggregate due to attractions between them and are converted into liquid. The same effect is produced by the increase of pressure. The gas molecules come closer by compression and coalesce to form the liquid.

Andres (1869) studied the P-T conditions of liquefaction of several gases. He established that for every gas there is a temperature below which the gas can be liquefied but above it the gas defies liquefaction. This temperature is called the critical temperature of the gas.

The *critical temperature*, T_c, of a gas may be defined as that temperature above which it cannot be liquefied no matter how great the pressure applied.

The *critical pressure*, P_c, is the minimum pressure required to liquefy the gas at its critical temperature.

The *critical volume*, V_c, is the volume occupied by a mole of the gas at the critical temperature and critical volume.

T_c, P_c and V_c are collectively called the *critical constants* of the gas. All real gases have characteristic critical constants.

At critical temperature and critical pressure, the gas becomes identical with its liquid and is said to be in *critical state*. The smooth merging of the gas with its liquid is referred to as the *critical phenomenon*.

Andrews demonstrated the critical phenomenon in gases by taking example of carbon dioxide.

ANDREWS ISOTHERMS OF CARBON DIOXIDE

The P-V curves of a gas at constant temperature are called *isotherms* or *isothermals*. For an ideal gas PV = nRT and the product PV is constant if T is fixed. Hence the isotherms would be rectangular parabolas.

For an ideal gas PV = nRT and the product PV is constant if π is fixed. Hence, the isotherms would be rectangular parabolas.

Andrews plotted the isotherms of carbon dioxide for a series of temperatures. There are three types of isotherms viz., those above 31°C, those below 31°C; and the one at 31°C.

(a) **Isotherms Above 31°C :** The isotherm at 25°C is a rectangular hyperbola and approximates to the isotherm of ideal gas. So are all other isotherms above 31°C. Thus, in the region above the isotherm at 31°C, carbon dioxide always exists in the gaseous state.

(b) **Isotherms Below 31°C :** The isotherms below 31°C are discontinuous. For example, the isotherm of 21°C consists of three parts.

   (i) *The Curve AB :* It is a PV curve for gaseous carbon dioxide. Along AB, the volume decreases gradually with the increase of pressure. At B the volume decreases suddenly due to the formation of liquid carbon dioxide having higher density.

   (ii) *The Horizontal Portion BC :* Along the horizontal part BC of the isotherm, the liquefaction continues while the pressure is held constant. At C all the gas is converted to liquid.

   (iii) *The Vertical Curve CD :* This part of the isotherm is, in fact, the P-V curve of liquid carbon dioxide. This is almost vertical since the liquid is not very compressible

METHODS OF LIQUEFACTION OF GASES

Various methods employed for the liquefaction of gases, depend on the technique used to attain low temperature. Now let us study Linde's Method and Claude's Method.

Linde's Method

Linde (1895) used Joule Thomson effect as the basis for the liquefaction of gases. *When a compressed gas is allowed to expand into vacuum or a region of low pressure, it produces intense cooling.* In a compressed gas the molecules are very close and the attractions between them are appreciable. As the gas expands, the molecules move apart. In doing so, the intermolecular attraction must be overcome. The energy for it is taken from the gas itself which is thereby cooled.

Linde used an apparatus worked on the above principle for the liquefaction of air. Pure dry air is compressed to about 200 atmospheres. It is passed through a pipe cooled by a refrigerating liquid such as ammonia. Here, the heat of compression is removed. The compressed air is then passed into a spiral pipe with a jet at the lower end. The free expansion of air at the jet results in a considerable drop of temperature. The cooled air which is now at about one atmosphere pressure passed up the expansion chamber. It further cools the incoming air of the spiral tube and returns to the compressor. By repeating the process of compression and expansion, a temperature low enough to liquefy air is reached. The liquefied air collects at the bottom of the expansion chamber.

Claude's Method

Claude's method for liquefaction of gases is more efficient than that of Linde. In this method also, the cooling is produced by free expansion of compressed gas.

The gas is made to do mechanical work by driving an engine. The energy for it comes from the gas itself which cools. *Thus in Claude's method the gas is cooled not only by overcoming the intermolecular forces but also by performance of work.* That is why the cooling produced is greater than in Linde's method.

The liquefaction or air. Pure dry air is compressed to about 200 atmospheres. It is led through a tube cooled by refrigerating liquid to remove any heat produced during the compression. The tube carrying the compressed air then enters the 'expansion chamber'. The tube bifurcates and a part of the air passes through the side-tube into the cylinder of an engine. Here it expands and pushes back the piston. Thus the air does mechanical work whereby it cools.

The air then enters the expansion chamber and cools the incoming compressed air through the spiral tube. The air undergoes further cooling by expansion at the jet and liquefies. The gas escaping liquefaction goes back to the compressor and the whole process is repeated over and over again.

ELECTRIC NATURE OF MOLECULES

Since molecules are made up of charged units, *i.e.*, electrons and the atomic nuclei, much of the behavior of molecules is understandable in terms of electrical interactions. A detailed theoretical treatment of a molecule by the methods of quantum mechanics can reveal the electron distribution of the molecule. This would allow the deduction of how the molecule would interact with other molecules, with a surrounding medium, or with an electric field. It is also possible to learn less detailed but useful information about the electric nature of molecules through an experimental approach. In this section the procedure which produces results for the *dipole moment* of a molecule is dealt with.

A voltage applied to a capacitor produces an electric field between the plates that depends on the capacitance of the capacitor.

Some basic ideas that enter into the treatment of charged particles in a vacuum (or approximately, in air) must be reviewed. This and the following section, where media other than vacuum are considered, provide the necessary background to the theory of the measurement of dielectric constants and the deduction of molecular dipole moments.

Electrostatic, or Coulombic, Forces

The basic relation in treating the interaction of stationary charges is Coulomb's law, which states that the force between two point charges is proportional to the product of their charges and inversely proportional to the square of the distance between them. The equality corresponding to this relation is written as

$$f = \frac{q_1 q_2 / (4\pi\varepsilon_0)}{r^2}$$

The force calculated from equation is obtained in newtons if the charges are expressed in coulombs, the separation of the charges is expressed in meters, and the proportionality factor $4\pi\varepsilon_0$ is given the value $1.11265 \times 10^{-10} C^2 N^{-1} m^{-2}$. For q_1 and q_2 of opposite signs the force is one of attraction; with the same signs, one of repulsion.

Electric Field

The interaction of charges at a distance suggests that an "electric field" exists around each charge. The intensity of the electric field at a point is defined as the force which would be exerted on a unit positive charge placed at that point. Thus, *the electric field strength E* and about a point charge q is the force on a unit charge, or

$$E = \frac{f}{q}$$

Let us consider an electric field about a point charge +q. At a distance r from this charge a unit positive charge would be repelled by a force

$$\frac{q/(4\pi\varepsilon_0)}{r^2}$$

The electric field strength at a distance r from the +q charge is therefore equal to

$$\frac{q/(4\pi\varepsilon_0)}{r^2}$$

Another important aspect of an electric field is described by the electric potential **V**. This quantity represents the potential energy of a unit positive charge in the electric field. A unit positive charge in an electric field of intensity **E** experiences a force equal to **E** . For a given displacement of the unit charge, the potential-energy change is the force, or the field strength, times the distance that the charge is moved. Since the potential energy increases as the unit positive test charge is brought closer to the positive charge q which generates the electric field, the change d**V** for an infinitesimal change dr is

$$d\mathbf{V} = \mathbf{E}\, dr$$

Rearrangement of equation gives a relation that is useful when electric field strengths are deduced from applied or known potentials. This expression is

$$E = \frac{dV}{dr}$$

The example of a plane-parallel capacitor with air (or, more properly, vacuum) between the plates can now be considered. If the capacitor is connected to a battery that produces a potential difference of **V** volts, there will be a potential drop of **V** volts across the capacitor. According to equation, the electric field E between the plates of the capacitor will be **V** /d, where d is the distance between the plates.

MOLAR POLARIZATION

The motor polarization of a substance depends on the polarizability and the dipole moment of the molecules of the substance.

A molecular explanation of the role of dielectric material in affecting electrical phenomena is now given. Our attention at first is restricted to the effect of the induced dipole moment that all molecules possess as a result of the electrical distortion of the electron distribution in a molecule by an applied electric field. All molecules are polarizable and therefore will have contributions from this factor.

The field **E**, acting on the molecule results from several contributions. The charges on the plates, the charges at boundaries of the dielectric adjacent to the plates, and the charges on the surface of a small cavity that is supposed to surround the molecule all contribute to the field on the molecule. The net result of these terms is

$$E_{int} = \frac{\sigma}{\varepsilon_0} - \frac{p}{\varepsilon_0} - \frac{p}{3\varepsilon_0} -$$

where the last term is the cavity-charge contribution. It is obtained by integrating, over the surface of the sphere, the effects of the surface charge in generating a field in the direction perpendicular to the capacitor plates. With the help of equation, one eliminates a and obtains

$$E_{int} = E + \frac{p}{3\varepsilon_0}$$

or $$p = 3\varepsilon_0(E_{int} - E)$$

The relation of p to $\mathbf{E}_{int}$ is obtained by using equation to eliminate **E**. This gives, after rearrangement,

$$p + \frac{3\varepsilon_0(\varepsilon / \varepsilon_0 - 1)}{\varepsilon / \varepsilon_0 + 2} = E_{int}$$

The proportionality between the polarization, which is the dipole induced in a unit volume of the dielectric, and the field acting on the molecules of the dielectric tells us about the electrical rigidity of these molecules. We interpret the polarization produced by the field E_{int} acting on the molecules in terms off the *molecular polarizability* α of these molecules. The number of molecules per unit volume is $(\rho/M)\mathbf{N}$, where ρ is the density and M is the molar mass. On this basis the relation between p and $\mathbf{E}_{int}$ is expressed as

$$p = \left[\left(\frac{\rho}{M}N\right)\alpha\right]E_{int}$$

Comparison with equation gives

$$\left(\frac{\rho}{M}N\right)\alpha + \frac{3\varepsilon_0(\varepsilon / \varepsilon_0 - 1)}{\varepsilon / \varepsilon_0 + 2}$$

It is customary to introduce the molar polarization 9^ by

$$P_M + \frac{\varepsilon/\varepsilon_0 - 1}{\varepsilon/\varepsilon_0 + 2}\ \frac{M}{\rho}$$

and then to recognize that the molecular polarizability a and the molar polarization $\mathbf{P}_M$ are related by

$$P_M = \frac{P\alpha}{3\varepsilon_0}$$

or

$$\alpha = \frac{3\varepsilon_0 P_M}{N}$$

We often use, $\alpha' \frac{\alpha}{(4\pi\varepsilon_0)}$ rather than α itself as the molecular polarizability then we get the equation

$$\alpha\ \frac{3P_M}{4\pi N}$$

MOLECULAR DIPOLE MOMENTS

Molecular dipole moments can be deduced from the temperature dependence of the molar polarization.

Consider now, as Peter Debye did in 1912, the contribution to the dielectric material of any permanent dipole moment μ of the molecules of the material. With no applied field the dipoles will be oriented in all directions and will be ineffective in contributing to the polarization **P** of the dielectric. In the presence of a field, however, the molecules tend to line up with the field so that their dipole moments add to the polarization **P**.

The energy of the dipole varies with the angle with which it is oriented to the acting field direction according to

$$\text{Energy} = \mu \mathbf{E}_{int} \cos\theta$$

The tendency of the molecules to go to the lowest-energy position by lining up with the field is opposed, however, by the thermal motions of the molecules. The distribution expression of Boltzmann gives the net effect of these two factors. Considerable manipulation is necessary for the calculation of the average dipole moment in the direction of the field.

Now let us study about the dipole moment from dielectric-constant measurements. The dielectric constant of a material is measured by placing it between the plates of a capacitor, or for liquids filling a cell in which the plates are inserted. This capacitor can be used as one arm of an electric bridge, like wheatstone bridge measuring resistances.

In principle the capacitance of the reference capacitor can be deduced from its geometry. In this way the capacitance of the sample capacitor can be obtained filled empty ad the electric constant of the sample can be deduced.

When the compound of interest is a liquid or solid material measurements are generally made on solutions of the material in some inert non polar substance, such as CCl_4 or benzene.

The Debye equation is based on the independent behavior of the polar molecules. Molecules with dipole moments exert considerable interaction on each other, and it is therefore best to apply the Debye equation to dilute solutions of polar compounds in nonpolar solvents. For gaseous samples the intermolecular distance is usually large enough for these interactions to present no difficulty.

Measurement of the dielectric constant and use of the Debye equation do not directly lead to the separate determination of the α and μ terms of equation. Two principal ways are available for sorting out these two factors.

In one procedure refractive-index data are used to obtain the molecular polarizability contribution to the molar polarizability. This contribution is subtracted from the molar polarizability calculated from dielectric-constant measurements, and any remaining molar polarization is attributed to the molecular dipole-moment contribution.

Another procedure consists of measuring ε and ρ as functions of temperature and using these data to plot $[(\varepsilon/\varepsilon_0 - 1)/(\varepsilon/\varepsilon_0 + 2)](M/\rho)$ versus $1/T$. The Debye equation leads us to expect such a plot to yield a straight line, the data for the hydrogen halides do behave in this manner.

From such plots the slope of the straight line can be used to obtain the dipole moment μ, and the intercept at $1/T$ can be made to yield the polarizability a. This procedure is quite straightforward and fails only if the molecules are associated to different extents at different temperatures or if the molecular configuration changes with temperature.

The order of magnitude of molecular dipole moments can be deduced by recognizing that these moments result from charges like that of an electron separated by molecular distances. For one electron separated from an equally charged positive center by a distance of 100 pm, the dipole moment would be

$$\mu = (1.6022 \times 10^{-19}\ \text{C})(100 \times 10^{-12}\ \text{m}) = 16.02 \times 10^{-30}\text{Cm}$$

In most of the chemical literature, molecular dipole moments are quoted in units of *debye*, a quantity based on the CGS-ESU system. The relation to SI units is

$$1\ \text{debye} = 3.338 \times 10^{-30}\text{Cm}$$

Thus, an electronic charge separated from one of opposite sign by a distance of 100 pm has a dipole moment of $(16.02 \times 10^{-30})(3.338 \times 10^{-30}) = 4.80$ debyes.

The information on molecular polarizabilities provided by these experiments finds rather less molecular-structure application than the dipole-moment results. This shows a qualitative correlation of molecular polarizability with the number of electrons in the molecule and the "looseness" with which they are bound.

Dipole Moments

For diatomic molecules the measured dipole moment of a molecule gives information about the displacement of the centre of negative charge from that of the positive charge along the internuclear direction. The assumption of that a dipole moment results from the location of the bonding electrons leads to a calculation of the ionic character from a dipole-moment.

MAGNETIC PROPERTIES OF MOLECULES

Magnetic measurements are a tool for molecular studies that have not been of such general applicability as electrical measurements. For certain types of compounds, however, magnetic measurements constitute one of the most powerful approaches to the elucidation of the arrangement of the electrons in the compound. The theory of magnetic studies parallels that of electrical studies so closely that a detailed treatment need not be given. Following some mention of the parallels between electrical and magnetic phenomena, the applications of magnetic studies are dealt with.

Magnetic susceptibility of substances, which can be determined experimentally, gives results for the magnetic polarizability and the magnetic dipole moment of the atoms, molecules, or ions of the substance. Magnets, iron filings for example, tend to be oriented in particular directions if they are near a conductor carrying an electric current. This effect leads us to say that an electric current is surrounded by a *magnetic field*. The strength and the direction of the field can be defined in terms of the effect it has on a test magnet, or a "magnetic dipole," at a particular location. The magnetic field strength, denoted by the symbol B, is called the *magnetic flux density*.

Guoy Balance

In the Guoy balance method, a sample is suspended from one arm of a balance in such a way that it is partly in the magnetic field. An electromagnet is ordinarily used, and when the magnet is turned on, the sample is generally repelled by or attracted into the magnetic field. The force required to maintain the position of the sample is measured by the weight that must be added or removed from the balance pan to maintain equilibrium. If the sample is paramagnetic, the magnetic moments will tend to line up with the field, and the sample will have lower energy in the magnetic field and will therefore be drawn into the field. If the sample is diamagnetic, the reverse will be the case, and the sample will be repelled by the field.

The relation between the force exerted on a sample by a nonhomogeneous magnetic field and the magnetic susceptibility of the sample can be determined. It is supposed that the sample is paramagnetic, so that the magnetic field lines up the microscopic magnets of the sample. The magnetic moment per unit volume that is affected by a magnetic flux density B is (κ/μ_0) B, and the magnetic moment of the section of thickness dz and volume A dz is therefore (κ/μ_0) B A dz.

To be specific, think of a paramagnetic material. Since the magnetic field increases along the sample toward the center of the magnetic field, the lowering of the potential energy of successive segments will be greater as a result of the lining up of the magnetic moments with the greater magnetic fields. The force corresponding to this varying potential energy is the rate of change of the potential with z. If the gradient of the magnetic flux density is dB/dz, the force on a sample segment is

$$\frac{\kappa}{\mu_0} \text{BA dz} \frac{dB}{dz} = \frac{\kappa}{\mu_0} \text{AB dB}$$

And if the segments of the sample extend from the center of the magnetic field, where the magnetic field value is B_0, to outside the magnetic field, where the value is zero, the total force in the sample is

$$f = \int_0^{B_0} \frac{\kappa}{\mu_0} AB\, dB = \frac{1}{2}\left(\frac{\kappa}{\mu_0}\right) AB_0^2$$

This result is the basis of the Gouy-balance method and shows that, for an experimental arrangement like, the measurement of the force exerted on the sample by a known maximum field strength B_0 can be used to deduce the magnetic susceptibility of the sample.

DIAMAGNETISM AND PARAMAGNETISM

All substances have a diamagnetic magnetic-polarizability component that acts to oppose an applied magnetic field.

All materials affect a magnetic field in which they are inserted as a result of an induced magnetic moment α_M, which produces a diamagnetic effect.

The diamagnetic effect is produced by the orbital motion of the electrons of the atoms, ions or molecules of the sample. The diamagnetic effect is temperature independent and depends on the nature of the electron orbitals.

Paramagnetism

Some normally nonmagnetic substances are attracted by a magnetic field, and studies of these "paramagnetic " substances give information about the number of unpaired electrons in the atoms, molecules, or ions of the substance.

The paramagnetic effect can be introduced most easily by considering the introductory problem of the magnetic behavior of an electron revolving about a nucleus. A classical treatment is easily made and yields a result which can then be converted to the correct quantum-mechanical result. The motion of an electron in an orbit corresponds, in this connection, to the passage of a current through a coil of wire. A current in a coil of wire of ordinary dimensions produces a magnetic field perpendicular to the coil. The magnetic field so produced is equal, according to Ampere's law, to that of a magnet with magnetic moment μ_M given by the product of the current and the cross-section area of the loop of wire.

$$\mu_M = iA$$

The current corresponding to an electron in orbit is obtained by multiplying the number of times the electron passes any point on the orbit by its electronic charge. Thus, with this classical picture of an electron in an atomic orbit, $i = [v/(2\pi r)]e$, where the electron velocity is v and the orbit has radius r. The cross- section area is $A = \pi r^2$. The magnetic moment μ_M is expressed as

This final form expresses the result of this classical derivation that can be carried over into quantum-mechanical systems, namely, that the ratio of the magnetic moment to the orbital angular momentum is equal to e/(2m).

The orbital angular momentum of an electron of an atom depends, on the quantum number l and is given by the expression $\sqrt{l(l+1)}\,h$. We can express the magnetic moment due to the orbital motion of the electron as

$$\mu_M = \hbar\sqrt{l(l+1)}\,(e/2m) = \frac{e\hbar}{2m}\sqrt{l(l+1)}$$

The constant factor in this equation provides a convenient unit in which to express the magnetic moment of atoms and molecules, and one therefore introduces the symbol μ_B, called, the electronic Bohr magneton as,

$$\mu_B = \frac{e\hbar}{2m}$$

With this unit, the orbital magnetic moment of an electron of an atom is given by

$$\mu_M = \mu_B\sqrt{l(l+1)}$$

When a similar approach is extended to molecules and ions, rather than free atoms, it would seem reasonable to expect the orbital motions of the electrons to contribute a magnetic moment of the order of an electronic Bohr magneton. This expectation is not generally borne out, and it appears that the orbital motions of the electrons in a polyatomic system are tied into the nuclear configuration of the molecule or the ion so tightly that they are unable to line up with the applied magnetic field and are therefore ineffective. Even for single-atom ions in solution, the interaction of the orbitals of the ion with the solvating molecules is apparently sufficient to prevent the orbitals from being oriented so that their magnetic moment contributes in the direction of the field. Thus, the orbital-magnetic-moment contribution to the magnetic susceptibility is generally quite small.

We must look to electron spin to explain the larger part of the magnetic moment of those molecules and ions which have magnetic moments. The association of a spin angular momentum of $\sqrt{s(s+1)}\hbar$, where s has the value of leads, according to equation, to the expectation of a spin magnetic moment. Atomic spectral data require, however, a magnetic moment that is twice that expected on the basis of the ratio of the magnetic moment to angular momentum implied by equation. Therefore, for the spin magnetic moment due to the electron spin, expressed in terms of the spin quantum number S of an atom or molecule, we have

$$\mu_M = 2\mu_B\sqrt{S(S+1)}$$

For one, two, three,... unpaired electrons, the spin-angular-momentum quantum number S is $\frac{1}{2}, \frac{2}{2}, \frac{2}{3}$, ... With equation and the assumption that the magnetic polarizability α_M contribution has been taken care of and that the orbital contribution to χ is negligible, the magnetic susceptibility of equation is related to the total electron spin by the relation

$$\chi = \left(\frac{4N\ \ \mu_0\mu_B^2}{3kT}\right)S(S+1) = \left(\frac{6.29\times10^{-6}}{T}\right)S(S+1)$$

At 25°C this expression gives

$$\chi = (2.11\times10^{-8})\ S\ (S+1)\ 25°C.$$

MOLECULAR ENERGIES

Translate rotational, and vibrational components of the energies fine molecules of a gas can be recognized.

A molecule is a collection of atoms held in a particular spatial arrangement by chemical bonds. For many purposes you can think of a molecule as a ball-and- spring system, with each atom being represented by a ball and each chemical bond by a spring. How do we describe the energy of such a system?

Think of having an actual ball-and-spring system to toss around. You would likely describe the energy of the system in terms of three types of motion:

Translation, consisting of the motion of the center of mass of the system *Rotation*, consisting of motions in which the system turns about one or more of the axes through the center of mass of the system.

Vibration in which the balls of the system oscillate about the equilibrium positions that correspond to the shape or structure of the system.

ENERGIES OF THE DIATOMIC MOLECULES OF A GAS

The analytical basis for treating these different types of motion can be seen by describing the motion of a diatomic molecule of a gas. Here we describe the potential-and kinetic-energy components of a freely moving gas-phase molecule treated as if it were a ball-and-spring system.

The only potential-energy contribution arises from variations in the distance between the atoms of the molecule. (The gravitational energy does not vary appreciably m any ordinary sample of a gas.) If the variable internuclear distance is represented by, then the potential energy U can be shown to be a function of r by writing U(r).

The kinetic energy depends on the motion of the two atoms of the molecule. We introduce the symbols $\dot{x}_1$, $\dot{y}_1$, and $\dot{z}_1$ to represent the velocity components dx_1/dt, dy_1/dt, and dz_1/dt, respectively, of one of the atoms. The symbols $\dot{x}_2$, $\dot{y}_2$, and $\dot{z}_2$ represent the velocity components of the second atom.

The total mechanical energy ε of the diatomic molecule is given by

$$e = \frac{1}{2}m_1(\dot{x}_1^2 + \dot{y}_1^2 + \dot{z}_1^2) + = \frac{1}{2}m_2(\dot{x}_2^2 + \dot{y}_2^2 + \dot{z}_2^2) + U(r)$$

Separate translational, rotational, and vibrational components of this energy can be recognized if a different coordinate system is introduced.

The de Broglie Wave Length

We can understand the behavior of beams of particles by using waves and the de Broglie wavelength.

In the early years of this century, physicists and chemists were confronted by a variety of puzzles as they investigated the atomic-molecular world. The prevalent attitude was that this world was populated by particles that behaved according to the laws of motion that hold for ordinary systems. But even the simplest atom, the hydrogen atom, could not be satisfactorily described. The line spectra of atomic hydrogen, and the spectra of other atoms, could not be related to the atoms responsible for these spectra. And along with such atomic puzzles was the awkwardly special wave-particle duality that had to be used to describe the behavior of electromagnetic radiation.

From the confusion of these years there emerged the recognition that " ordinary," or classical, mechanics could not be used to treat some of the behavior of atomic-world particles. A new method for describing the behavior of particles, a new "mechanics," had to be developed.

One special characteristic of this mechanical behavior is that generally only certain energies are "allowed." The energy is said to be restricted to certain values, or to be *quantized*. The "mechanics" that we use in which situations is called *quantum mechanics*.

In quantum mechanics we use wave pictures or, mathematical expressions called wave functions to arrive at the allowed behaviors of atomic or molecular particles. As a result, the term *wave mechanics* is also used.

To describe the translational, rotational, and vibrational motions of molecules, we now use the methods of quantum, or wave, mechanics. These molecular motions can be simply treated, and they provide a good introduction to the general methods of quantum mechanics and to the nature of the quantized world. First the background to our acceptance of wave mechanics is sketched.

Vibrational Energies

In a classical, *i.e.*, nonquantum mechanical description, the atoms of the molecules are pictured as moving back and forth relative to one another as the molecule vibrates. The region in which the atoms move as a result of this vibrational motion is a small fraction, typically about one-tenth, of the equilibrium bond length. Thus, the atoms are confined to a region in space and just as for the "square-well" system the quantum mechanical treatment will show that only certain energies are allowed. These allowed vibrational energies can be deduced by applying the Schrödinger equation.

CLASSICAL VIBRATIONS

The vibration of the atoms of a molecule relative to one another is best introduced by considering first a classical ball-and-spring system. Consider a ball, or particle, that is attached to a spring. The spring exerts a restoring force, pulling or pushing it back to its equilibrium position. Suppose the force is proportional to the distance to which the particle has been displaced from that position. The vibrational motion of both ordinary-sized objects held by actual springs and atoms held by chemical bonds can be treated with this simplest-spring description.

A spring that exerts a restoring force proportional to the displacement from the equilibrium length is said to *obey Hooke's law*. Since displacing the particle in one direction brings about a force in the opposite direction, Hooke's law is written

$$f = -kx$$

where f is the restoring force and x is the displacement from the equilibrium position. The proportionally constant k is known as the *force constant* and is a measure of the stiffness of the spring. The force constant is equal to the restoring force operating for a unit displacement from the equilibrium position.

The classical motion of a particle can be deduced from Newton's law f = ma. If f = –kx and a = dx^2/dt^2 are substituted, one obtains

$$m\frac{d^2x}{dt^2} = -kx$$

or

$$\frac{m}{k}\frac{d^2x}{dt^2} = -x$$

A solution to this equation can be seen by inspection and verified by substitution to be

$$x = A \sin \sqrt{\frac{k}{m}t}$$

Here A is a constant that is equal to the maximum value of x; that is, it is the vibrational amplitude.

The position of the particle varies sinusoidally with time. Each time t increases by $2\pi\sqrt{m/k}$, the quantity $\sqrt{k/mt}$ increases by 2π and the particle traces one complete cycle. The time corresponding to one oscillation, or vibration, is therefore $2\pi\sqrt{m/k}$. More directly useful is the reciprocal of this quantity, which is the frequency of vibration *i.e.*, the number of vibrations performed per second.

If this quantity is denoted,

V_{vib}, then we have $V_{vib} = \frac{1}{2\pi}\sqrt{\frac{k}{m}}$.

For a system of ordinary dimension this natural frequency of oscillation depends on the values of k, and m. Any amount of energy, can be imparted to the vibrating system, and this energy changes only the amplitude of the vibration.

ANGULAR MOMENTUM

Two different types of motion can be recognized from the treatments of molecular motions. One is longitudinal. Translational and vibrational motions are of this type. The other is rotational. Molecular rotations, and the "spin" of particles such as electrons, are examples of this type of motion. Both types of motion can be treated by the Schrödinger-equation procedure. For rotary motions, the characteristics of the allowed states can more conveniently be found by considering the angular momentum of the rotating system.

Angular momentum of the electron of a hydrogen atom was the central feature in the historic 1913 treatment of the hydrogen atom by Niels Bohr. Bohr's recognition of quantized angular momentum anticipated the general results that can be seen from the Schrödinger equation. Bohr's method provides a basis for the deduction of the allowed energies of all types of rotating particles and systems of particles of the molecular world. As an introduction to this route to quantized states, we consider some features of macroscopic, or classical, rotations.

Classical Rotations

Before we see the implications of the restrictions placed on angular momentum, some of the quantities, such as angular momentum itself, that are used for rotating systems are introduced.

Consider a particle rotating about a fixed point. The speed of rotation can be described in terms of the particle's velocity υ or by the number of revolutions per second, *i.e.*, the frequency ν. Since the particle travels a distance υ each second and the distance for one revolution is $2\pi r$, we have the relation

$$\nu = \frac{\upsilon}{2\pi r}$$

The *angular velocity* ω is the number of radians (rad) swept through per second. Since there are 2π rad in one revolution, $\omega = 2\pi\nu$, and

$$\omega = \frac{\upsilon}{r}$$

For a rotating, many-particle, rigid system, the angular velocity ω is particularly convenient. Whereas each particle moves with a velocity υ that depends on its distance r from the axis of rotation, all particles move around this axis with the same value of ω.

The kinetic energy of a particle, $1/2m\upsilon^2$, can be rearranged to a form more convenient for rotary systems. We first rewrite $1/2m\upsilon^2$ as

$$\varepsilon = \frac{1}{2}mr^2\left(\frac{\upsilon}{r}\right)^2 = \frac{1}{2}(mr^2)(\omega^2)$$

The term mr^2 is a one-particle example of a frequently occurring collection of terms. It is convenient to introduce I, the *moment of inertia*, for mr^2, where m is the mass of the particle and r is its distance from the axis of rotation.

If $I = mr^2$ then we get

$$\varepsilon = \frac{1}{2} I\omega^2$$

This expression has been developed for a single particle rotating about a fixed point. It is also applicable to a system of particles, such as a molecule, rotating about the center of mass. Then the moment of inertia is defined by $I = \Sigma_i m_i r_i^2$, where m_i is the mass of the *i*th particle and r_i is its distance from the center of mass and the summation is over all the particles of the rotating system.

The magnitude of the angular momentum of a single particle rotating in a circle is given by the expression $m\upsilon r$, where $m\upsilon$ is the linear momentum and r is the distance from the center of rotation. This angular-momentum expression is extended to many-particle systems by introducing the moment of inertia. We recognize that

$$\text{Angular momentum} = m\upsilon r = (mr^2)\left(\frac{\upsilon}{r}\right) = I\omega.$$

With $I = \Sigma_i m_i r_i^2$ this expression can be applied to systems of particles rotating about a fixed point. The rotational energy is expressed in terms of the angular momentum by

$$\varepsilon = \frac{1}{2} I\omega^2 \left(\frac{I\omega}{2I}\right)^2$$

ROTATIONAL ENERGIES

As we know the rotational motions of a gas molecule are quantized.

The rotational motion of linear molecules is easiest to treat. Such molecules undergo a relatively simple end-over-end rotational motion.

The integers, or quantum numbers, that index the allowed total angular momenta of a rotating molecule are usually represented by the symbol J. Thus, the allowed angular momenta of the rotating molecule are

$$\sqrt{J(J+1)}\hbar \qquad J = 0, 1, 2, \ldots$$

The allowed rotational kinetic energies ε_J,

$$\varepsilon_J = J(J+1)\frac{\hbar^2}{2I} \quad J = 0, 1, 2, \ldots$$

Each rotational energy level, corresponding to a particular value of J, consists of $2J + 1$ states. This is the degeneracy of the rotational-energy levels, and with the symbol g_J for this degeneracy we write

$$g_J = 2J + 1$$

A general diagram showing the pattern of allowed rotational energies of a linear molecule is given. Generally, these spectral studies make use of radiation in the microwave spectral region.

A rotational-energy-level diagram for a particular molecule, complete with an energy scale, can be constructed, if a value of the moment of inertia of the molecule is available.

Spectral studies provide such data. Also, the diffraction studies provide molecular-structure data from which moment-of-inertia values can be calculated. Thus, the information needed to calculate the energies which any gas-phase molecule has as a result of its rotational motion is generally available.

THREE-DIMENSIONAL TRANSLATIONAL ENERGY

The three-dimensional partition function and its derivative with respect to temperature can be used to reach an expression for the thermal energy of three-dimensional translational motion. The result, obtained is

$$(U - U_0)\text{trans} = \frac{3}{2}RT$$

Although, q_{trans} depends on the mass of the particles and the volume of the container, the thermal energy $(U - U_0)$trans has for 1 mol of any gas in any volume, the value $\frac{3}{2}RT$

Distribution Over Quantum States

The distribution expressions for three-dimensional motion can be derived by following the same procedure as we did for one-dimensional motion. First, however, we see that we can use one "effective" quantum number n in place of the three quantum numbers n_x, n_y, and n_z.

he allowed energies are given by the expression,

$$n_x^2 + n_y^2 + n_z^2 + \left(\frac{h^2}{8ma^2}\right)$$

It is enough for us to deal with a quantity that shows the sum of the squares of the quantum numbers rather than with the individual values. We introduce the variable n defined by $n^2 = n_x^2 + n_y^2 + n_z^2$. Then the allowed energies are given by $n^2h^2/(8ma^2)$ instead of the more detailed, but no more useful, expression involving n_x, n_y, and n_z.

In using the effective quantum number n, we must recognize that there are a number of states all with the same value of n, or of energy ε_n. The number of states at this energy is the degeneracy g_n. The display of states as points for large n, the additional number of states included when n increases by 1 is equal to $1/2\pi n^2$. Thus, if we use n as an effective quantum number, we must use $g_n = 1/2\pi n^2$ as the degeneracy. The distribution expressions will now follow, rearranged to

$$\frac{N_n}{N} = \frac{g_n}{q} e^{-(\varepsilon_n - \varepsilon_0)/(kT)}$$

For the three-dimensional translational motion that we are dealing with, the following specific relations apply

$$\varepsilon_n = \frac{n^2h^2}{8ma^2} \varepsilon_n = \frac{n^2h^2}{8ma^2}$$

$$g_n = \frac{1}{2}\pi n^2$$

$$q = \left(\frac{2\pi mkT}{h^2}\right)^{3/2} V$$

Substitution of these three specific expressions in the general distribution expression, and rearrangement, gives

$$\frac{1}{N} = \frac{dN}{dn} = \frac{1}{N} N_n = \frac{\pi n^2}{2}\left(\frac{h^2}{2\pi mkT}\right)^{3/2} \frac{1}{V} e^{-n^2(h^2/(8ma^2))/(kT)}$$

We use this cumbersome distribution-over-n expression only to learn about the distribution over energies and speeds.

2

Radioactivity

INTRODUCTION

A nuclear reaction, it is the nucleus of the atom which is involved. Now let us study about Radioactivity and types of Radiations. A number of elements such as uranium and radium are unstable. Their atomic nucleus breaks of its own accord to form a smaller atomic nucleus of another element. The protons and neutrons in the unstable nucleus regroup to give the new nucleus. This causes the release of excess particles and energy from the original nucleus, which we call *radiation*. The elements whose atomic nucleus emits radiation are said to be *radioactive*. The spontaneous breaking down of the unstable atoms is termed *radioactive disintegration* or *radioactive decay*.

The disintegration or decay of unstable atoms accompanied by emission of radiation is called Radioactivity. The radiations are divided into three types. In 1902 these rays were sorted out by Rutherford. The rays which bend towards the negative plate, which carry positive charge, were named α-rays. Those bending towards the positive plate arid carrying negative charge were called β (*beta*) *rays*. The third type of radiation, being uncharged, passed straight through the electric field and were named γ (*gamma*) *rays*. α, β and γ rays could be easily detected as they cause luminescence on the zinc sulphide screen placed in their path.

USE OF THE RADIOACTIVITY

Radioactive Tracers

A very important use of radioactivity, which is fast covering an ever-widening field of applications in many branches of science, is what is known as *radioactive indicators* or *isotopic tracers*. It is based on the existence of radioactive elements which are isotopic with stable elements;

e.g., RaB, RaD, ThB and AcB are all isotopic with stable lead; RaC, RaE, ThC and AcC are likewise isotopes of bismuth. Now, if RaD, one of the radioactive isotopes of lead, is mixed with the latter, it behaves like an indicator or tracer and the lead can be traced by means of the associated RaD, even if the lead were to be found in such a small quantity that it cannot be investigated by ordinary methods. The tracer remaining chemically inseparable and indistinguishable from the element to be traced will always give information on the behaviour of the other, following it through all, even complex, physical and chemical changes. With the discovery of induced radioactivity and the production of radioactive isotopes by artificial means, the importance of the isotopic tracers has been greatly enhanced.

In the purely physical and chemical fields, indicators have been used in the determination of the solubility of sparingly soluble materials, measurement of rates of diffusion in solid systems, study of phenomena at boundary surfaces and of intermolecular exchanges, etc. *Medical applications* also have been made. For instance, in the treatment of syphilis, by employing RaE as indicator, the retention of bismuth in the organism has been investigated in detail and it has been found that for a long time after treatment a considerable amount of bismuth remains in the body, maintaining an antisyphilitic effect. But the major application of isotopic tracers at present is in *biological researches* where artificially produced radioelements are used with ease and economy. With their aid, the rate, place and sequence of formation of the organic constituents of a living body, the permeability of cell-walls, the metabolism of phosphorus in human, animal aud plant systems, the rate of lymph circulation, etc., have been investigated. *Some of the other interesting applications are:*

Luminescent Paints

A carefully prepared mixture of radio-thorium (α-emitter) with zinc sulphide exhibits a more or less permanent luminescence and is used for coating the pointers and figures of clocks and watches, for rendering visible signs in theatres and so on.

Radium Therapy

The rays from radium, though harmful to healthy skin, produce satisfactory improvement in various skin diseases. They also have a healing effect on internal diseases. The curative properties of many natural springs have been attributed to their content of radon,

Had nature not dispersed the radioactive substances through the rocks of the earth, had there not been one or two of them long-lived enough to survive and maintain a supply of their descendants until man arrived and became scientific, or if the faint outward signs of radioactivity latent in the rocks had been overlooked or once noticed had been left unstudied, in any of these cases centuries more might have passed before a proper foundation was laid for the edifice of the atom. That is the prime reason for honouring those who discovered radioactivity and they did not rest until they hod brought it fully into the light.

Their is an illustrious history, but not without pathos. And the benefits which they gave have not yet been fully exploited; marvellous things may still be discovered in the process of understanding the action of the rays on living matter.

NUCLEAR ENERGY

A heavy isotope as Uranium-235 (or plutonium-239) can undergo nuclear chain reaction yielding vast amounts of energy. *The energy released by the fission of nuclei is called nuclear fission energy or nuclear energy*. Sometimes, it is incorrectly referred to as *atomic energy*.

The fission of U-235 or Pu-239 occurs instantaneously, producing incomprehensible quantities of energy in the form of heat and radiation. If the reaction is uncontrolled, it is accompanied by explosive violence and can be used in atomic bombs. However, when controlled in a reactor, the fission of U-235 is harnessed to produce electricity.

Nuclear Chain Reaction

We know that U-235 nucleus when hit by a neutron undergoes the reaction,

$$^{235}_{92}U + ^{1}_{0}n \rightarrow ^{139}_{56}Ba + ^{94}_{36}Kr + ^{1}_{0}n.$$

Each of the three neutrons produced in the reaction strikes another U-235 nucleus, thus causing nine subsequent reactions. These nine reactions, in turn, further give rise to twenty seven reactions. This process of propagation of the reaction by multiplication in threes at each fission, is referred to as a *chain reaction*. Heavy unstable isotopes, in general, exhibit a chain reaction by release of two or three neutrons at each fission. It may be defined as:

A fission reaction where the neutrons from a previous step continue to propagate and repeat the reaction.

A chain reaction continues till most of the original nuclei in the given sample are fissioned. However, it may be noted that not all the neutrons released in the reaction are used up in propagating the chain reaction. Some of these are lost to the surroundings. Thus for a chain reaction to occur, the sample of the fissionable material should be large enough to capture the neutron internally. If the sample is too small, most neutrons will escape from its surface, thereby breaking the chain. *The minimum mass of fissionable material required to sustain a chain reaction is called critical mass.* The critical mass varies for each reaction. For U-235 fission reaction it is about 10 kg.

As already stated, even a single fission reaction produces a large amount of energy. A chain reaction that consists of innumerable fission reactions will, therefore, generate many times greater

The chain reaction of Uranium-235 produces heavy amount of energy. Now let us study about nuclear energy.

Atomic Bomb

A bomb which works on the principle of a fast nuclear chain reaction is referred to as the atomic bomb.

It contains two subcritical masses of fissionable material,

^{235}U or ^{239}Pu.

It has a mass of trinitrotoluene in a separate pocket.

When TNT is detonated, it drives one mass of ^{235}U into the other. A supercritical mass of the fissionable material is obtained. As a result of the instantaneous chain reaction, the bomb explodes with the release of tremendous heat energy.

The temperature developed in an atomic bomb is believed to be 10 million °C (temperature of the sun). Besides many radionuclei and heat, deadly gamma rays are released.

These play havoc with life and environment. If the bomb explodes near the ground, it raises tons of dust into the air. The radioactive material adhering to dust is known as *fall out.* It spreads over wide areas and is a lingering source of radioactive hazard for long periods.

IMPORTANT METHOD OF RADIOACTIVITY

The important methods to detect the radioactive radiation can be studied as follows :

Cloud Chamber

The chamber contains air saturated with water vapour. When the piston is lowered suddenly, the gas expands and is super cooled. As an α- or β-particle passes through the gas, ions are created along its path. These ions provide nuclei upon which droplets of water condense. The trial produced marks the track of the particle. The Cloud Chamber Method gives better photographs.

Ionisation Chamber

This is the simplest device used to measure the strength of radiation. An ionisation chamber is fitted with two metal plates separated by air. When radiation passes through this chamber, it knocks electrons from gas molecules and positive ions are formed. The freed electrons migrate to the anode and positive ions to the cathode.

Thus, a small current passes between the plates. This current can be measured with an ammeter, and gives the strength of radiation that passes through the ionisation chamber. In an ionisation chamber called *Dosimeter*, the total amount of electric charge passing between the plates in a given time is measured. This is proportional to the total amount of radiation that has gone through the chamber. To measure the radioactivity modern scintillation counter can be used effectively. Scintillation counter can measure radiation up to a million per second. Now let us study about Film Badges. A film badge consists of a photographic film encased in a plastic holder. When exposed to radiation, they darken the grains of silver in photographic film. The film is developed and viewed under a powerful microscope.

As α- or β-particles pass through the film, they leave a track of black particles. These particles can be counted. In this way the type of radiation and its intensity can be known. However, γ-radiation darken the photographic film uniformly. The amount of darkening tells the quantity of radiation

A film badge is an important device to monitor the extent of exposure of persons working in the vicinity of radiation. The badge-film is developed periodically to see if any significant dose of radiation has been absorbed by the wearer.

PROPERTIES OF RADIATIONS

Alpha (α), beta (β) and gamma (γ) rays differ from each other in nature and properties. Their chief properties are:

(a) Velocity;

(b) Penetrating Power;

(c) Ionisation.

Alpha-Rays

Nature : They consist of streams of α-particles. *By measurement of their e/m, Rutherford showed that they have a mass of 4 amu and charge of +2.* They are *helium nuclei* and may be represented as $^{4}_{2}\alpha$ or $^{4}_{2}He$.

Velocity

α-particles are ejected from radioactive nuclei with very high velocity, about one-tenth that of light.

Penetrating Power

Because of their charge and relatively large size, α-particles have *very little power of penetration* through matter. They are stopped by a sheet of paper, 0.01 mm thick aluminium foil or a few centimetres of air.

Ionisation

They cause *intense ionisation* of a gas through which they pass. On account of their high velocity and attraction for electrons, α-particles break away electrons from gas molecules and convert them to positive ions.

Beta-Rays

Beta rays are the streams of β-particles emitted by the nucleus.

They have very small mass. (1/1827 amu.)

β-rays travel 10 times faster than α-particles. Their velocity is about the same as of light.

β-rays are 100 times more penetrating in comparison to α-particles. Because they have higher velocity and negligible mass.

The ionisation produced by β-particles in a gas is about one-hundredth of that of α-particles. These rays are poor ionisers.

Gamma-Rays

Unlike α-and β-rays, they do not consist of particles of matter. They have no mass or charge; they travel with the velocity of light.

Their ionising power is very weak, in comparison to, α-and β-particles. Since their photon electron collisions are small, γ-rays are weak ionisers.

NUCLEAR REACTIONS

In a chemical reaction there is merely a rearrangement of extranuclear electrons. The atomic nucleus remains intact. A nuclear reaction involves a change in the composition of the nucleus. The number of protons in the nucleus is altered. The product is a new nucleus of another atom with a different atomic number and/or mass number. Thus,

The nuclear reaction is one which proceeds with a change in the composition of the nucleus so as to produce an atom of a new element.

The conversion of one element to another by a, nuclear change is called *transmutation.*

We have already consider the nuclear reactions of radioactive nuclei, producing new isotopes. Here we will consider such reactions caused by artificial means.

Differences Between Nuclear Reactions and Chemical Reactions

Some characteristic differences between nuclear reactions and ordinary chemical reactions are listed below:

Nuclear Reactions	Chemical Reactions
1. Proceed by redistribution of nuclear particles.	1. Proceed by the rearrangement of extranuclear electrons.
2. One element may be converted into another.	2. No new element can be produced.
3. Often accompanied by release or absorption of enormous amount of energy.	3. Accompanied by release or absorption of relatively small amount of energy.
4. Rate of reaction is unaffected by external factors such as concentration, temperature, pressure and catalyst.	4. Rate of reaction is influenced by external factors.

NUCLEAR FISSION REACTIONS

In these reactions an atomic nucleus is broken or fissioned into two or more fragments. This is accomplished by bombarding an atom by alpha particles ($_2^4He$), neutrons ($_0^1n$), protons ($_1^1H$), deutrons ($_1^2He$), etc. All the positively charged particles are accelerated to high kinetic energies by a device such as a *cyclotron*. This does not apply to neutrons which are electrically neutral. The projectile enters the nucleus and produces an unstable '*compound nucleus*'. It decomposes instantaneously to give the products. For example, $_7^{14}N$ when struck by an α-particle first forms an intermediate unstable compound nucleus, $_8^{18}F$, which at once cleaves to form stable $_8^{17}O$.

NUCLEAR FUSION REACTIONS

These reactions take place by combination or fusion of two small nuclei into a larger nucleus. At extremely high temperatures the kinetic energy of these nuclei overweighs the electrical repulsions between them. Thus, they coalesce to give an unstable mass which decomposes to give a stable large nucleus and a small particle as proton, neutron, positron, etc. For example : Two hydrogen nuclei, $_1^1H$, fuse to produce a deuterium nucleus, $_1^2H$. Deuterium nucleus, $_1^2H$, and tritium nucleus, $_1^3H$, combine to give helium nucleus, $_2^4He$ with the expulsion of a neutron. Nuclear Fission and Nuclear Fusion have got many differences:

Nuclear Fission	Nuclear Fusion
1. A bigger (heavier nucleus splits into smaller (lighter) nuclei.	1. Lighter nuclei fuse together to form the heavier nucleus.
2. It does not require high temperature	2. Extremely high temperature is required for fusion to take place.
3. A chain reaction sets in.	3. It is not a chain reaction.
4. It can be controlled and energy released can be used for peaceful purposes.	4. It cannot be controlled and energy released cannot be used properly.
5. The products of the reaction are radioactive in nature.	5. The products of a fusion reaction are non-radioactive in nature.
6. At the end of the reaction nuclear waste is left behind.	6. No nuclear waste is left at the end of fusion reaction.

Nuclear Equations

Similar to a chemical reaction, nuclear reactions can be represented by equations. *These equations involving the nuclei of the reactants and products are called nuclear equations.* The nuclear reactions occur by redistribution of protons and neutrons present in the reactants so as to form the products.

Thus, the total number of protons and neutrons in the reactants and products is the same. Obviously, *the sum of the mass numbers and atomic numbers on the two sides of the equation must be equal.*

If the mass numbers and atomic numbers of all but one of the atoms or particles in a nuclear reaction are known, the unknown particle can be identified.

While writing nuclear equations certain steps have to be followed. They are as follows:

1. Write the symbols of the nuclei and particles including the mass numbers (superscripts) and atomic numbers (subscripts) on the left (reactants) and right (products) of the arrow.
2. Balance the equation so that the sum of the mass numbers and atomic numbers of the particles (including the unknown) on the two sides of the equation are equal. Thus, find the atomic number and mass number of the unknown atom, if any.
3. Then look at the periodic table and identify the unknown atom whose atomic number is disclosed by the balanced equation.

Nuclear equations can written as follows:

Disintigration of radium –236 by emission of an alpha particle ($^{4}_{2}He$).

$$^{236}_{88}Ra \rightarrow {}^{232}_{86}Rn + {}^{4}_{2}He.$$

NUCLEAR REACTORS

Nuclear reactor is a specially designed plant, where controlled fission is carried.

The chief components of a nuclear reactor are as follows:

1. *U-235 fuel rods* which constitute the 'fuel core'. The fission of U-235 produces heat energy and neutrons that start the chain reaction.

2. *Moderator* which slows down or moderates the neutrons.

   The most commonly used moderator is ordinary water. Graphite rods are sometimes used. Neutrons slow down by losing energy due to collisions with atoms/molecules of the moderator.

3. *Control rods* which control the rate of fission of U-235. These are made of boron-10 or cadmium, that absorbs some of the slowed neutrons.

$$^{10}_{5}B + ^{1}_{0}n \rightarrow ^{7}_{3}Li + ^{4}_{2}He.$$

Thus the chain reaction is prevented from going too fast.

4. *Coolant* which cools the fuel core by removing heat produced by fission. Water used in the reactor serves both as moderator and coolant. Heavy water (H_2O) is even more efficient than light water.
5. *Concrete shield* which protects the operating personnel and environments from destruction in case of beakage of radiation.

Light-Water Nuclear Power Plant

Most commercial power plants today are 'light-water reactors'. In this type of reactor, U-235 fuel rods are submerged in water. Here, water acts as coolant and moderator. The control rods of boron-10 are inserted or removed automatically from spaces in between the fuel rods.

The heat emitted by fission of U-235 in the fuel core is absorbed by the coolant. The heated coolant (water at 300°C) then goes to the *exchanger*. Here the coolant transfers heat to sea water which is converted into steam. The steam then turns the turbines, generating electricity. *A reactor once started can continue to function and supply power for generations.*

1. The fusion fuel, deuterium ($^{2}_{1}H$), can be obtained in abundance from heavy water present in sea water. The supplies of U-235 needed for a fission reactor are limited.
2. A fusion reaction produces considerably greater energy per gram of fuel than a fission reaction.
3. The products of fusion ($^{3}_{2}He$, $^{4}_{2}He$) are not radioactive. Thus, there will be no problem of wast disposal.

So far, it has not been possible to set up a fusion reactor. The chief difficulty is that the reactant nuclei must be heated to very high temperatures. A mixture of deuterium and tritium nuclei, for example, requires 30 million °C before they can fuse. So far no substance is known which can make a container that could withstand such high temperatures. However, scientists are making efforts to effect fusion at a lower temperature with the help of laser beams.

Hydrogen Bomb

This deadly device makes use of the nuclear fusion of the isotopes of hydrogen. It consists of a small plutonium fission bomb with a container of isotopes of hydrogen. While the exact reaction used is a strictly guarded military secret, a fusion reaction between H-2 and H-3 may be the possible source of the tremendous energy released.

$$^{2}_{1}H + ^{3}_{1}H \rightarrow ^{4}_{2}H + ^{1}_{0}n + [\text{tremendous energy}]$$

The 'fusion bomb' produces the high temperature required for nuclear fusion and triggers the H-bomb.

The explosion of such a bomb is much more powerful than that of a fission bomb or atomic bomb. Fortunately, the H-bombs have been tested and not used in actual warfare. If they are ever used, it may mean the end of civilisation on earth.

MOLE FRACTION

A solution is a homogeneous mixture of two or more substances on molecular level constituent of the mixture present in a smaller amount is called the Solute and the one present in a larger amount is called the *Solvent.*

For example, when a small amount of sugar (solute) is mixed with water (solvent), a homogeneous solution in water is obtained. In this solution, sugar molecules are uniformly dispersed in molecules of water. Similarly, a solution of salt (Na^+ Cl) in water consists of ions of salt (Na', CF) dispersed in water.

The concentration of a solution is defined as :

The amount of solute present in a given amount of solution.

Concentration is generally expressed as the quantity of solute in a unit volume of solution.

$$\text{Concentration} = \frac{\text{quantity of solute}}{\text{Volume of solution}}$$

A solution containing a relatively low Concentration of solute is called *Dilute solution.* A solution of high concentration is called *Concentrated solution.*

We can identity seven types of solutions. They are as follows :

State of Solute	State of Solvent	Example
1. Gas	Gas	Air
2. Gas	Liquid	Oxygen in water, CO_2 in water (Carbonated drinks)
3. Gas	Solid	Adsorption/of H_2 by palladium
4. Liquid	Liquid	Alcohol in water
5. Liquid	Solid	Mercury in silver
6. Solid	Liquid	Sugar, Salt
7. Solid	Solid	Metal alloys; Carbon in iron (Steel)

In fact, substances in any three states of matter (solid, liquid, gas) can act as solute or solvent. Thus there are *seven types of solutions.*

Of the seven types of solutions, we will discuss only the important ones in detail.

There are several ways of expressing concentration of a solution :

(a) Per cent by Weight (b) Mole Fraction

(c) Molarity (d) Molality

(e) Normality.

Percent by Weight

It is the *weight of the solute as a per cent of the total weight of the solution.* That is,

$$\% \text{ by weight of solute} = \frac{\text{Wt. of solute}}{\text{Wt. of solution}}$$

For example, if a solution of HCl contains 36 per cent HCl by weight, it has 36 g of HCl for 100 g of solution.

Example:

What is the per cent by weight of NaCl if 1.75 g of NaCl is dissolved in 5.85 g of water.

Solution:

Wt. of solute (NaCl) = 1.75g

Wt. of solvent (H_2O) = 5.85 g

Wt. of solution $= 1.75 + 5.85 = 7.60$ g

Hence concentration of NaCl % by weight

$$= \frac{1.75}{7.60} \times 100$$

$$= \mathbf{23.0}.$$

Mole Fraction

A simple solution is made of two substances : one is the solute and the other solvent Mole traction, X, of solute is defined as *the ratio of the number of moles of solute and the total number of moles of solute and solvent.* Thus,

$$X_{solute} = \frac{\text{moles of solute}}{\text{moles of solute} + \text{moles of solvent}}$$

If n represents moles of solute and N number of moles of solvent,

$$X_{solute} = \frac{n}{n+N}$$

Notice that mole fraction of solvent would be

$$X_{solvent} = \frac{n}{n+N}$$

Mole fraction is unitless and

$$X_{solute} + X_{solvent} = 1.$$

HENRY'S LAW

As we know the solubility of a gas in a solvent depends on the pressure and the temperature. We can notice the following equilibrium, when a gas is enclosed over its saturated solution.

$$\text{gas} \rightleftharpoons \text{gas in solution.}$$

The pressure can be reduced, by more gas dissolving in solvent. Thus solubility or concentration of a gas in a given solvent is increased with increase of pressure.

In 1803 William Henry investigated the relationship between pressure and solubility of a gas in a particular solvent. His generalisation was known as Henry's law. The statement is as follows :

For a gas in contact with a solvent at constant temperature, concentration of the gas that dissolves in the solvent is directly proportional to the pressure of the gas.

Mathematically, Henry's Law may be expressed as

$$C \propto P$$
$$C = k\,P$$

where P = pressure of the gas; C = concentration of the gas in solution; and k = proportionality constant known as *Henry's Law Constant*. The value of k depends on the nature of the gas and solvent, and the units of P and C used.

Henry's law contains certain limitations, which applies closely to gases with nearly ideal behavior.

These limitations can be marked as follows :

1. At moderate temperature and pressure.
2. If the solubility of the gas in the solvent is low.
3. The gas does not react with the solvent to form a new species. Thus, ammonia (or HCl) which react with water do not obey Henry's Law.

$$NH_3 + H_2O \rightleftharpoons NH_4^+ + OH^-$$

4. The gas does not associate or dissociate on dissolving in the solvent

Example:

The solubility of pure oxygen in water at 20°C and 1.00 atmosphere pressure is 1.38×10^{-2} mole/litre. Calculate the concentration of oxygen at 20°C and partial pressure of 0.21 atmosphere.

Solution:

Calculation of k:

$$C = kP$$

or $$k = \frac{C}{P}$$

Substituting values for pure oxygen,

$$k = \frac{1.38 \times 10^{-3}\,\text{mole / litre}}{1.00\text{ atm}} = 1.38 \times 10^{-3}\,\frac{\text{mol / litre}}{\text{atm}}$$

and C = kP

$$\text{concentration of } O_2 = \frac{1.38 \times 10^{-3}\,\text{mole / litre}}{\text{atm}} \times 0.21\text{ atm.}$$
$$= 2.0 \times 10^{-4}\text{ mole/litre.}$$

SOLUTIONS OF LIQUIDS

The solutions of liquids in liquids may be divided into three classes.

1. Liquids that are completely miscible (*e.g.*, alcohol and water).
2. Liquids that are partially miscible (*e.g.*, ether and water).
3. Liquids that are partially immiscible (*e.g.*, benzene and water).

Liquids like alcohol and ether mix in all properties, and in this respect they could be compared to gases.

A large number of liquids are known which dissolve in one another only to limited external, ether and water. Ether dissolves, about 1,2% water and water also dissolves about 6.5% ether. Since their mutual solubilities are limited, they are only partially miscible. When equal volumes of ether and water are shaken together, two layers are formed one of a saturated solution of ether in water, and the other a saturated solution of water in ether.

These two solutions are referred to as conjugate solutions. The effect of temperature on toe mutual solubility of these mixtures of conjugate solutions is of special interest. We will study the effect of temperature on the composition of such mixtures with reference to three typical systems:

(i) Phenol-Water System

(ii) Triethylamine-Water System

(iii) Nicotine-Water System.

Now let us study about, phenol-water system.

Phenol-Water System

The miscibility of phenol and water. The left hand side of the parabolic curve represents one of the two conjugate solutions which depicts the percentage of phenol dissolved in water at various temperatures. The solubility of phenol increases with temperature. The right hand side of the curve represents the other conjugate solution layer that gives the percentage of water in phenol. The solubility of water in phenol also increase with increase of temperature. The two solution curves meet at the maxima on the temperature-composition curve of the system. This point here corresponds to temperature 66°C and composition of phenol as 33%. Thus, at a certain maximum temperature the two conjugate solutions merge, become identical and only one layer results. The temperature at which the two conjugate solutions (or layers) merge into one another to form one layer, is called the *Critical Solution Temperature.*

The determination of critical solution temperature may, therefore, be used for testing the purity of phenol and other such substances.

At any temperature above the critical solution temperature, phenol and water are miscible in all proportions. Outside the curve there is complete homogeneity of the system, *i.e.,* one layer only exists; and under the curve there may be complete miscibility but it depends upon the composition of the mixture. It is clear that at a temperature below 50°C a mixture of 90% phenol and 10% water or 5% phenol and 95% water, will be completely miscible since the corresponding points do not lie under the curve. Two layers will always separate out below the curve and the curve gives compositions of the two conjugate solutions constituting the two layers. At 50°C a mixture of equal proportion of phenol and water (50% each) will form two-layers whose compositions are given by A and B. The line joining the potato (M and N) corresponding to the compositions A and B is called the *tie line.* This line helps in calculating the relative amounts of the two layers, which is here-given by the ratio *MN/ML.*

Some other liquid pairs behaving like phenol-water system are given below with their CST values and the percentage of the first component being given in bracket

(a) Methanol-Cyclohexane (49°C; 29)

(b) Hexane-Aniline (59.6°C; 52)

(c) Carbon Disulphide-Methanol (49.5°C; 80),

Now let us study about the mutual solubilities of triethylamine and water.

Triethylamine-Water System

Unlike phenol-water system, the solubilities decrease with the increase in temperature in this system. The two conjugate solutions mix up completely at or below 18.5°C. This temperature is also called the critical solution temperature or the lower consolate temperature. As in the above case, any point above the horizontal line corresponds to heterogeneity of the system (two layers) while below it is complete homogeneity (one layer). Thus an equi-component mixture (50-50) will be completely miscible at 10°C but at 50°C there will be separating out two layers having compositions corresponding to the points C and D.

Common examples of this system with their lower critical solution temperatures and percentage of the first component are given below.

(a) Diethylamine-Water (43°C; 13)

(b) U-Methylpiperidine-Water (48°C; 5).

The combination of nicotine and water is also more or less same as the previous two systems.

Nicotine-Water System

At ordinary temperature nicotine and water are completely miscible but at a higher temperature the mutual solubility decreases and as the temperature is raised further the two liquids again become miscible. In other words, the mutual solubility increases both on lowering as well as raising the temperature in certain ranges. Thus, we have a closed solubility curve and the system has two critical-solution-temperatures, the upper 208°C and the lower 61°C. The effect of pressure on this system is that the lower critical temperature is raised while the upper critical temperature is lowered gradually until finally they become one. At this point the liquids are miscible at all the temperatures.

STEAM DISTILLATION

Distillation which is carried in a current of steam is called steam distillation. This technique is widely used for purification of organic liquids, which are steam volatile and immiscible with water.

The impure organic liquid admixed with water containing nonvolatile impurities is heated and steam passed into it. The vapour of the organic liquid and steam rising from the boiling mixture pass into the condenser. The distillate collected in the receiver consists of two layers, one of the pure organic liquid and the other of water. The pure liquid layer is removed by means of a separatory funnel and further purified.

The vapour pressure of a liquid rises with increase of temperature. When the vapour pressure equals the atmospheric pressure, the temperature recorded is the boiling point of the given liquid. In case of a mixture of two immiscible liquids, each component exert its own vapour pressure as if it were alone. The total vapour pressure over the mixture (P) is equal to the sum of the individual vapour pressures (p_1, p_2) at that temperature.

$$P = p_1 + p_2$$

Hence, the mixture will boil at a temperature when the combined vapour pressure P, equals the atmospheric pressure. Since $P > p_1$ or p_2 *the boiling point of the mixture of two liquids will by lower than either of the pure components.*

In steam distillation the organic liquid is mixed with water (bp 100°C). Therefore the organic liquid will boil at a temperature lower than 100°C. For example, phenylamine (aniline) boils at 184°C but the steam distillation temperature of aniline is 98°C.

Steam distillation is particularly used for the purification of an organic liquid (such as phenylamine) which decomposes at the boiling point and ordinary distillation is not possible. Relative amounts of Organic liquid and Water distilling over.

The number of molecules of each component in the vapour will be proportional to its vapour pressure *i.e.*, to the vapour pressure of the pure liquid at that temperature.

Hence,
$$\frac{n_1}{n_2} = \frac{p_1}{p_2}$$
where n_1 and n_2 are the number of moles of the two components in the vapour.

Now, from (1), we have

$$\frac{w_1 / M_1}{w_2 / M_2} = \frac{p_1}{p_2}$$

where w_1 and w_2 are the masses of the two liquids distilling over, and M_1 and M_2 their molecular weights.

Since one of the two components is water (m.wt. = 18), we can write from (2)

$$\frac{\text{mass of organic liquid}}{\text{mass of water}} = \frac{p_1 \times M_1}{p_2 \times 18}$$

Thus, the ratio of masses of the organic liquid and water can be calculated from the given values of p_1, p_2 (aqueous tension) and M_1, the molecular weight of the organic liquid.

DISSOLUTION AND RECRYSTALLISATION

When a solid is placed in a solvent, molecules or ions, as the case may be, break away from the surface and pass into the solvent. The particles of the solid thus detached are free to diffuse throughout the solvent to give a uniform solution. The solute and the solvent molecules are constantly moving about in the solution phase because of kinetic energy possessed by them. Some of the particles are deflected back towards the solid on account of collisions with other molecules. These then strike the solid surface and may get entangled in its crystal lattice and thus get deposited on it. This process by which the solute particles from solution are 'redeposited' or 'recrystallised' is often spoken of as *recrystallisation* or *precipitation*.

In a solution in contract with solid solute, therefore, two opposing processes are operating simultaneously:

(a) Dissolution

The particles of the solute leaving the solid and passing into solution.

(b) Recrystallisation

The particles of the solute returning from the solution and depositing (or precipitating) on the solid.

To start with the rate at which the particles leave the solid is much greater than the rate at which they return to it. As the number of particles of the solute in solution increases, the rate at which they are returned to the solid also increases. Eventually, if there is excess of solid present the rate of dissolution and the rate of recrystallisation become equal. At this stage, a state of equilibrium between the molecules of the solute in solution and the solid solute is said to have been reached. Thus,

$$\text{Solute solid} \rightleftharpoons \text{Solute dissolved}$$

Henceforth neither the amount of the solute in solution nor the solid phase present in contact with it, will change with lapse of time. This equilibrium state will remain so, provided the 'kinetic energy' of the molecules is not changed by a change in temperature.

The competition between the two processes and the eventual equally of the rates of the two processes, points to an important phenomenon in chemistry called the *dynamic equilibrium*. The term dynamic refers to the fact that both the processes are occurring continuously but due to the equality of the two rates (equilibrium) no net change in the amount of the solute in solution phase occurs with the passage of tune.

The dynamic nature of solubility equilibrium can be demonstrated by putting a crystal of sugar having a hole in the surface, in a solution of sugar which has already attained the state of equilibrium. It will be seen that after some time the hole is filled with solid sugar and the concentration ot the solution remains unchanged. This is explained by saying that the molecules from the solution get settled in the hole (process of crystallisation) while some molecules from other parts of the crystal go into the solution (process of dissolution) thereby changing the shape of the crystal.

In general, when a solid solute is in dynamic equilibrium, with its solution, the rate of dissolution (R_d) evidently depends upon the number of molecules leaving the crystal surface. The larger the area or liquid crystal surface the greater will be the rate of dissolution. That is represented as follows:

$$R_d \propto A$$

or $$R_d = k_d \times A \qquad ...(1)$$

where k_d may be called the dissolution constant. Its value is characteristic of a particular system and its value depends on temperature.

The rate of recrystallisation (R_r) is the rate at which the solute molecules return to the crystal surface from solution and are deposited on it This is determined by two factors:

(a) The surface (A) of the crystal; the larger the area, the greater the number of molecules settling on it;

(b) The concentration C of the solute molecules in solution; the higher the number of solute molecules in solution, the greater their number settling down. Thus,

$$R_r \propto A \times C$$

or $$R_r = k_r \times A \times C \qquad ...(2)$$

where k_r may be called the recrystallisation constant. Its value is also characteristic of a system and depends upon temperature.

At equilibrium the rate of dissolution and the rate of recrystallisation are equal. $R_d = R_r$.

Or, from (1) and (2) above

$$k_d \times A = k_r \times A \times C$$

$$C = \frac{k_d}{k_r} \text{ constant}$$

$$= K \text{ (say)}.$$

Hence, the concentration of solute at equilibrium state in the solution is constant for a particular solvent and at a fixed temperature. The solution thus obtained is called a 'Saturated solution' of the solid substance and the concentration of this solution is termed its 'Solubility'.

Thus, a saturated solution is defined as one which is in equilibrium with the excess of solid at a particular temperature.

The solubility is defined as the concentration of the solute in solution when it is in equilibrium with the solid substance at a particular temperature.

Each substance has a characteristic solubility in a given solvent. The solubility of a substance is often expressed in terms of number of grams of it that can be dissolved in 100 grams of the solvent.

For example, a saturated solution of sodium chloride in water at 0°C contains 35.7 g of NaCl in 100 g of H_2O. Or. The solubility of NaCl in water at 0°C is 35.7 g/100 g.

When a saturated solution prepared at a higher temperature is cooled, it gives a solution which would contain more solute than the saturated solution at that temperature. Such a solution is called a *supersaturated solution.* Supersaturated solutions are quite unstable and change to the saturated solution when excess of solute precipitates out. This fact is utilized in the purification of chemical substances

Raoult's Law

As a result of extensive experimentation, Raoult (1886) gave an empirical relation connecting the relative lowering of vapour pressure and the concentration of the solute in solution. This is now refereed to as the *Raoult's Law*. It states that :

the relative lowering of the pressure of a dilute solution is equal to the mole fraction of the solute present in dilute solution.

Raoult's Law can be expressed mathematically in the form:

$$\frac{p - p_s}{p} = \frac{n}{n + N}$$

n = number of moles or molecules of solute.

N = number of moles or molecules of solvent.

Dilute solutions containing non-volatile solutes exhibit the following properties :

1. Lowering of the Vapour Pressure,

2. Elevation of the Boiling Point,
3. Depression of the Freezing Point,
4. Osmotic Pressure.

The essential feature of these properties is that they depend only on the number of solute particles present in solution. Being closely related to each other through a common explanation, these have been grouped together under the class name *Colligative Properties* (Greek colligatus = collected together).

A colligative property may be defined as one which depends on the number of particles in solution and not in any way on the size or chemical nature of the particles.

Consequent to the above definition, each colligative property is exactly related to any other. Thus if one property is measured, the other can be calculated. The colligative properties of dilute solutions are particularly important as these provide valuable methods for finding the molecular weights of the dissolved substances.

The vapour pressure of a pure solvent is decreased when a non-volatile solute is dissolved in it. If p is the vapour pressure of the solvent and p_s that of the solution, the lowering of vapour pressure is $(p - p_s)$. This lowering of vapour pressure relative to the vapour pressure of the pure solvent is termed the Relative lowering of Vapour pressure. Thus,

$$\text{Relative Lowering of Vapour Pressure} = \frac{p - p_s}{p}.$$

Now let us think about the derivation of Raoult's law.

The vapour pressure of the pure solvent is caused by the number of molecules evaporating from its surface. When non volatile solute is dissolved in solution, the presence of solute molecules in the surface blocks a fraction of the surface where no evaporation can take place. This causes the lowering of the vapour pressure. The vapour pressure of the solution is, therefore, determined by the number of molecules of the solvent present at any time in the surface which is proportional to the mole fraction. That is,

$$p_s \propto \frac{N}{n + N}$$

where N = moles of solvent and n = moles of solute.

or
$$p_s = k\frac{N}{n + N}$$

k being proportionality factor.

In case of pure solvent n = 0 and hence

Mole fraction of solvent $= \frac{N}{n+N} = \frac{N}{0+N} = 1$

Now from equation (1), the vapour pressure p = k

Therefore the equation (1) assumes the form

$$p_s = p\frac{N}{n+N}$$

$$\frac{p_s}{p} = \frac{N}{n+N}$$

$$1 - \frac{p_s}{p} = 1 - \frac{N}{n+N}$$

$$\frac{p - p_s}{p} = \frac{n}{n+N}$$

This is Raoult's law.

ENERGIES OF ATOMS AND MOLECULES

A molecule or an atom or an electron exists in one of the allowed states and has the energy corresponding to that state. Collections of large numbers of molecules, like those one deals with in ordinary-sized, or *macroscopic*, systems, contain molecules distributed throughout the allowed states. Many of the properties of chemical materials can be deduced if the energies of the quantum states and the distribution throughout these states are known. In practice, the deduction of properties of macroscopic samples from such detailed calculations is often impossibly difficult. But the calculations can be carried out for many ideal-gas systems, and the general principles of the procedure can be appreciated.

Any ordinary sample of a gas contains very many molecules. Now that we know about the possible energies of individual molecules, how do we describe the energy of a large collection of molecules? That question is a specific example of the heart of chemistry—the explanation of macroscopic behavior in terms of atomic-molecular world behavior.

We need to use only the general ideas that were developed about the allowed energies of molecules.

The energies of molecules can be described in terms of the energy in each degree of freedom : The motions of a molecule of a gas can be thought of in terms of motion in each "degree of freedom."

We must know the allowed energies for the translational motion of a molecule along one coordinate, one of the three translational degrees of freedom. This expression applies to each of the three translational degrees of freedom of a gas molecule.

The rotation of a linear molecule can be described in terms of rotation around the two axes that can be drawn perpendicular to the internuclear axis of the molecule. A linear molecule has 2 rotational degrees of freedom. (Since no molecule has a single rotational degree of freedom, the allowed rotational energies of a linear molecule with its 2 rotational degrees of freedom.) The motion of a generally shaped molecule cannot, it turns out, be described in terms of a simple combination of rotations around three orthogonal axes only a general guide to the allowed energies for the 3 rotational degrees of freedom of a generally shaped molecule.

Quantum Numbers

The allowed energies for each type of motion are indexed by a quantum number. As you see the expressions for these numbers enter in various ways into the expressions for the allowed energies. These expressions, and therefore the form in which the quantum number "index" enters, result from the mathematical treatment that we use to get the expression for the allowed energies.

Degeneracy

Generally, several quantum states occur with the same energy and, therefore, are implied by the quantum number for that energy level. Recall, for example, that the allowed rotational energy of a linear molecule is specified by the value of J, the rotational quantum number. There are, however, 2J + 1 different rotational states with this energy. The degeneracy of the Jth rotational energy level is 2J + 1. The symbol g is used for the degeneracy.

Boltzmann Distribution

Boltzmann distribution takes very important role in chemistry. As we know, the particular distribution of molecules throughout the set can be studied by following equation.

$$W\begin{pmatrix} N_1 & N_2 & N_3 & \\ g_1 & g_2 & g_3 & \cdots \end{pmatrix} = \frac{N!}{N_1!N_2!N_3!\ldots}$$

We want to investigate the values of W for various distributions, subject to these constraints:

1. The total number of particles is fixed; *i.e.*, we are dealing with a particular sample.
2. The system of particles has some fixed total energy.

The way in which W varies with the type of distribution can be seen by dealing with specific simple systems.

Consider an energy pattern in which there is only one state at each of the allowed energies. Then all the g, values equal 1. This simplifies the expression for the number of ways in which the distribution can be achieved to

$$W = \frac{N!}{N_1!N_2!N_3!\ldots}$$

To begin, we deal with a manageable number of particles instead of Avogadro's number. Then the N, values are small enough that the value of W can be calculated directly. From the W values for various distributions—subject to these two constraints—the distribution that corresponds to the largest value of W, that is, the most probable distribution, can be discovered.

An artificial, but informative. Notice that in all cases the total number of particles and the total, or average, energy are the same.

Then notice that the more the particles are spread out through the available states, subject to the two constraints, the larger the value of W. The largest value of W is obtained with the final distribution. The most probable distribution is that in which the particles are spread out as much as possible. The bunching up in the lower-energy states is more than compensated for by a spreading out through the higher-energy states.

From the figure we can understand more clearly about the number of particles per energy level. The way the most probable distribution varies with the total energy of the system, or the average energy of the particles of the system, is illustrated by the examples. All the distributions have the same general shape, but the lower the average energy, the more the particles are crowded into the lower-energy levels. The higher the average energy, the more the particles are spread out into the higher-energy levels.

From the preceding qualitative considerations, you see that the most probable distribution of molecules throughout the energies allowed to them depends on the average energy. From the kinetic-molecular studies you saw that the average energy, or the energy per mole of particles. depends on the absolute temperature. It follows that the most probable distribution depends on the absolute temperature. The *Boltzmann distribution* shows this dependence for most systems that are of interest in chemistry.

The Boltzmann distribution shows the number of particles *per state* at one energy compared to that at another energy.

PARTITION FUNCTION

The Boltzmann distribution expression gives directly the population of the states at one energy compared to the population of the states at another energy. This expression can be used when, for example, we are interested in the population of high-energy "excited" states. More often we are interested in the properties of the entire system of particles.

To deal with the macroscopic behavior of systems of particles, we must use the Boltzmann distribution expression to find out how many of the particles of the system behave in each of the allowed ways. The Boltzmann distribution equation must be applied to each of the very many states available to the particles. This can be done in a very elegant and revealing way by introducing a particular summation over all the available states. This summation is known as the *partition function.*

The partition function, given the symbol q, is a summation that weights the quantum states in terms of their availability and then adds the resulting terms.

The partition function is defined by

$$q = \sum g_i \, e^{(\varepsilon_i - \varepsilon_0)/(kT)}.$$

The partition function turns out to be a very convenient single quantity that can be used to express the properties of a system of many particles spread out over the states available to them.

Notice that the value of q can be calculated for any particular type of atomic or molecular motion if the energy pattern of the allowed states is known. Often, as for the translational, rotational, and vibrational energies of gas-phase molecules, the pattern is known. Then a numerical value of q can be obtained.

In other cases the specific information on the energies of the allowed states is not known. Even then the ideas suggested by the partition function are of value.

The Boltzmann distribution relates the number of particles N_i in any energy level to, for example, the number No in the lowest-energy level. We often want to deal with the total number of particles rather than with the number in the lowest energy level. The partition function helps us do this. We start with the Boltzmann distribution of

$$\frac{N_i}{g_i} = \left(\frac{N_o}{g_o}\right) e^{-(\varepsilon_i - e_o)/(kT)}$$

The total number of particles is given by the summation

$$N = \sum N_i = \frac{N_o}{g_o} \sum g_i \, e^{-(\varepsilon_i - e_o)/(kT)}$$

You can recognize the summation as what we have called the partition function q. Thus, we have

$$N = \frac{N_o}{g_o} q$$

or

$$\frac{N}{q} = \frac{N_o}{g_o}$$

The second expression suggests that the entire set of particles have q states available to them, just as the No molecules have the g_o states available to them.

Now we can return to the general expression for N_i/g_i given by and replace N_o/g_o by N/q. This gives us the form of the Boltzmann distribution that we will find most useful

$$\frac{N_i}{g_i} = \left(\frac{N}{q}\right) e^{-(\varepsilon_i - e_o)/(kT)}$$

This expression lets us calculate the population N_i/g_i of any state, or the population N, of any energy level, from the total number of particles.

In many connections we deal with the energy of a chemical system. Often we think of two contributions to this energy. One is the energy that the system would have if all the particles were in the lowest-energy states. The second is the *thermal energy*, the energy that the system has because the particles are distributed throughout some of the higher-energy states. We use the symbol U for the energy of 1 mol of particles. The symbol U_0 denotes the energy that the system would have if all the particles were in the lowest-energy states. It follows that $U - U_0$ is the thermal energy.

The thermal energy is given by the summation

$$U - U_0 = N_o(0) + N_1(\varepsilon_1 - \varepsilon_0) + (\varepsilon_2 - \varepsilon_0) + \ldots$$
$$= \Sigma N(\varepsilon_1 - \varepsilon_0).$$

This summation could be developed by using the Boltzmann term for each N_i. The result what we get will be as follows:

From

$$q = \Sigma g_i e^{-(\varepsilon_i - \varepsilon_0)/(kT)}$$

we get,

$$\frac{dq}{dT} = \Sigma g_i e^{-(\varepsilon_i - \varepsilon_0)/(kT)} \frac{\varepsilon_i - \varepsilon_0}{kT^2}$$

This can be modified to,

$$kT^2 \frac{dq}{dT} = \Sigma[g_i e^{-(\varepsilon_i - \varepsilon_0)/(kT)}(\varepsilon_i - \varepsilon_0)]$$

$$= \Sigma\left[\left(\frac{g_i}{N_i}\right)e^{-(\varepsilon_i - \varepsilon_0)/(kT)} N_i(\varepsilon_i - \varepsilon_0)\right].$$

This equation can be written as follows:

$$kT^2 \frac{dq}{dT} = \frac{q}{N} \Sigma N_i(\varepsilon_i - \varepsilon_0)$$

$$= \frac{q}{N}(U - U_0)$$

We get,

$$(U - U_0) = \frac{RT^2}{q}\frac{dq}{dT}$$

or

$$(U - U_0) = RT^2 \frac{d(\ln q)}{dT}.$$

The above equation shows that for any type of motion, the thermal energy can be calculated if an expression which can be differentiated with respect to temperature is obtained for

$$q = \Sigma g_i e^{-(\varepsilon_i - \varepsilon_0)/(kT)}.$$

ROTATIONAL AND VIBRATIONAL MOTIONS

An integration can be used to obtain the rotational partition function and the rotational contribution to the thermal energy.

As the molecules of a gas fly around in all directions, they tumble end over end. This rotational motion is quantized, and the energies of the rotational states are already studied. Now we ask about the net consequence of these allowed rotational states on the affairs of a collection, say 1 mol, of gas molecules.

Rotational Partition Function—Linear Molecules

The allowed energies and degeneracies of a rotating linear molecule, with its 2 rotational degrees of freedom, are given by

$$\varepsilon_J = J(J+1)\frac{\hbar^2}{2I}$$

and $$g_J = 2J + 1 \qquad \text{where } J = 1, 2, \ldots$$

From these expressions for energies and degeneracies an expression for the rotational partition function can be developed.

We begin with

$$q_{rot} = \sum_{J=0}^{\infty}(2J+1)e^{-J(J+1)[\hbar^2/(2I]/(kT)}$$

For most molecules at not too low a temperature, the values of J that will lead to most of the contributing terms in this summation are very large compared with unity. We can write

$$q_{rot} = \sum_{J=0}^{\infty} 2Je^{-J^2\hbar^2/(2IkT)}$$

Furthermore, the large number of summation terms that contribute allows us to treat J as a continuous variable, so that the summation can be replaced by an integral. Thus we have

$$q_{rot} = 2\int_{J=0}^{\infty} Je^{-J^2\hbar^2/(2IkT)}dI$$

Use of the appropriate integral from Appendix Sec. A-I allows this to be evaluated to give

$$q_{rot} = \frac{kT}{\hbar^2/(2I)} = \frac{2IkT}{\hbar^2}$$

For a generally shaped molecule, rotation can be described in terms of component rotations about three perpendicular axes. The moments of inertia about these axes are known as the principal moments of inertia and are represented by I_A, I_B, and I_c. Since now the pattern of allowed energies is not so easily expressed, the partition function is cited without derivation.

For a generally shaped molecule with identical atoms that can be interchanged by a rotation of the molecule, a symmetry number appears in the denominator. The symmetry number is again the number of different ways the molecule can be turned to give back an original position. The symmetry number of H_2O is 2, of NH_3 is 3, and of CH_4 is 12.

The rotational partition function for a generally shaped molecule is

$$q_{rot} = \left(\frac{2IkT}{\hbar^2}\right)^{3/2} \frac{(\pi I_A I_B I_C)^{1/2}}{\sigma}$$

The general relation to the 2-degrees-of-freedom equation is apparent.

The rotational contribution to the thermal energy is again obtained. The result is the expected expression 3/2RT.

A summation can be used to obtain the vibrational partition function and the vibrational contribution to the thermal energy.

With the classical picture, gas molecules fly around, tumble end over end, and vibrate. The atoms of a diatomic molecule are described as vibrating against one another, as do two balls joined by a spring. But the vibrational energy is quantized, and the expression for the allowed vibrational energies. Now we see how the molecules of a macroscopic sample are distributed throughout these allowed vibrational states.

Vibrational Partition Function

Each way in which a molecule can vibrate leads to a set of vibrational states with energies given by

$$\varepsilon_{\text{vib mode}} = \left(\upsilon + \frac{1}{2}\right)\frac{h}{2\pi}\sqrt{\frac{k}{m}} = \left(v + \frac{1}{2}\right)h\nu_{vib} \qquad \upsilon = 0,\ 1,\ 2,\ .$$

For each vibrational mode we can thus write

$$q_{\text{vib mode}} = \sum_{\upsilon=0}^{\infty} e^{-[(\upsilon+1/2)h\nu_{vib}-(1/2)h\nu_{vib}](kT)} \sum_{\upsilon=0}^{\infty} e^{-\upsilon h\nu_{vib}/(kT)}$$

Now, in contrast to translational and rotational energy spacings, the vibrational spacings are appreciable compared with kT, and therefore only a few of the terms in the series will contribute appreciably to the partition function. The partition function sum cannot be replaced by an integral, but we can obtain an expression by developing the summation

$$q_{vib} = 1 + e^{-h\nu_{vib}/(kT)} + e^{-2h\nu_{vib}/(kT)} +$$

With the introduction of the convenient symbol

$$x = \frac{h\nu_{vib}}{kT}$$

this becomes

$$q_{vib\ mode} = 1 + e^{-x} + e^{-2x} + \ldots$$
$$= 1 + (e^{-x}) + (e^{-x})^2 + \ldots$$

This series can be recognized as the binomial expansion of $(1-e^{-x})^{-1}$ and thus, we have

$$q_{vib\ mode} = \frac{1}{1-e^{-x}} = \frac{1}{1-e^{-h\nu_{vib}(kT)}}$$

HEAT CAPACITIES

Here we anticipate one of the properties of substances that are introduced then. The *heat capacity* is a measure of the energy required to raise the temperature of a one-mole sample of a substance by one kelvin, or one degree Celsius.

You will see that heat capacities can be measured by suitable calorimetric methods. In most of these studies, the pressure is kept constant, as it is if the calorimeter is a container open to the atmosphere. These results can be corrected, or other more difficult experiments can be carried out, to give values for the heat capacity of the sample when its volume does not change. The heat capacity at constant volume C_V can be most simply understood. The heat capacity at constant pressure C_P has in it the complexities that result when energy is used to push back the surroundings when the temperature increases and the sample expands.

Think of 1 mol of a gas in a container of fixed volume. With your understanding of the energies of the molecules of a gas, you can understand the constant- volume heat capacity. The mathematical expression for the heat capacity is

$$C_v = \frac{dU}{dT}$$

The energy U of a sample, a 1-mol sample of a gas in this case, can be expressed as the sum of the thermal energy $U - U_0$ and the energy U_0 that the sample would have if only the lowest-energy states were populated. When we deal with heat capacities, we need to deal with only the temperature-dependent, thermal energy $U - U_0$.

Molecular Contributions to the Heat Capacity of 1 mol of an Ideal Gas ($x = h\nu_{vib}/(kT) = 1.438\bar{\nu}/T$, $\bar{\nu}$ in cm^{-1}).

Translation	$C_{trans} = \frac{d}{dT}\left(\frac{3}{2}RT\right) = \frac{3}{2}R$
Rotation	$C_{rot} = \frac{d}{dT}\left(\frac{2}{3}RT\right) = R$
Linear molecule	$C_{rot} = \frac{d}{dT}\left(\frac{3}{2}RT\right) = \frac{3}{2}R$
Nonlinear molecule	
Vibration	$C_{vib} = \frac{d}{dT}\left(\frac{N\ h\nu_{vib}}{e^{h\nu_{vib}/(kT)} - 1}\right) = \frac{3x^2e^x}{(e^x - 1)^2}$ (per degree of freedom)

The thermal energy for most cases can be treated in terms of the three contributions given by

$$U - U_0 = (U - U_0)_{trans} + (U - U_0)_{rot} + (U - U_0)_{vib}$$

It follows that the heat capacity C_v can be thought of in terms of the contributions given by

$$C_v = C_{trans} + C_{rot} + C_{vib}.$$

Each heat-capacity contribution can be calculated from the derivative with respect to temperature of the expression for the corresponding thermal energy. Expressions for the translational, rotational, and vibrational heat capacity contributions are shown in Table.

Vibrational Heat Capacity

Only the C_{vib} term requires a treatment that is specific for the gas beings studied. The derivative of the vibrational-energy expression of equation with respect to temperature gives

$$C_{vib} = kN \left(\frac{h\nu_{vib}}{kT}\right) = \frac{e^{h\nu_{vib}/(kT)}}{(e^{h\nu_{vib}/(kT)} - 1)^2}$$

$$= R\frac{x^2e^x}{(e^x - 1)^2}$$

[Use has been made of the expressions R = **N** k and $x = h\nu_{vib}/(kT)$.]

The contribution of a molecular vibrational mode to the heat capacity can be calculated for any vibrational-energy spacing $h\nu_{vib}$ and any temperature, or for any value of $x = h\nu_{vib}/(kT)$.

For many purposes the graphical treatment of the C_{vib} function, provides adequate values. The heat-capacity contribution of a vibrational mode as a function of temperature for various vibrational-energy spacings. Notice that for any spacing the heat-capacity contribution approaches the classical value of R at high temperatures.

The following table indicates the calculation of the heat capacity C_v of SO_2 at 25°C.

Contributing Term		Heat-Capacity Contribution, $JK^{-1}\ mol^{-1}$	
Translation		$\frac{3}{2}R = 12.47$	
Rotation		$\frac{3}{2}R = 12.47$	
Vibration :			
$\bar{\nu} = 518\ cm^{-1}$	$x = 2.50$	$R\frac{x^2e^x}{(e^x - 1)^2} = 5.06$	
$\bar{\nu} = 1151\ cm^{-1}$	$x = 5.56$	1.00	
$\bar{\nu} = 1362\ cm^{-1}$	$x = 6.58$	0.50	6.56
			31.50

HARDNESS OF CRYSTALS

The vibrational-energy spacing factor depends, on the force constants of the bonds between the atoms or the ions and on the mass of these particles. The Einstein or Debye energy spacing factors are also proportional to these quantities. We can write

$$v_{max} \propto \sqrt{\frac{k}{m}} \text{ or } k \propto mv^2_{max}$$

As studied, we can use the value of mv^2_{max} to arrange simple crystals in order of the force constant which describes the strength of the bond between adjacent particles. The graduation from soft, weakly bonded crystals to hard, strongly bonded ones is evident.

Liquids

Now the C_v data of liquids can be considered. Most informative, initially, is the value of just about 3R tor mercury. The simplest explanation of this result comes from considering that the mercury atoms of the liquid have, not 3 translational degrees of freedom, which would lead to a heat capacity of only 3/2 but 3 vibrational degrees of freedom. If the atoms are only loosely bound to their neighbours and the vibrational frequencies are low, each vibrational contribution will be equal to R, and the total contribution will be 3R.

With the assumption of a contribution of 3R from the liquid-state counterpart of translational motion, we can proceed to polyatomic molecules. The new feature that enters is the counterpart of the molecular rotation. The contribution from the rotational-type motion, however, is not easily handled. If the motion were entirely free, there would be contributions of 1/2R per degree of freedom.

Calculations of the Heat Capacity C_v of Some Liquids at 25°C JK^{-1} mol^{-1} (Calculations for free and restricted rotation)

Motion of center	Hg		CS_2		CCl_4	
of mass	3R = 24.9		3R = 24.9		3R = 24.9	
Intramolecular vibrational motion (see Sees. 4-8 and 4-9)	0		10.3		45.1	
Rotationlike motion	Free	Restricted	Free	Restricted	Free	Restricted
	0	0	R=8.3	2R=16.6	3/2R=12.5	3R=24.9
C_v, calculated	24.9	24.9	43.5	51.8	82.5	94.9
Observed	23.6		47.1		89.5	

However, the interference of neighbouring molecules is such that this motion is more like a low-frequency vibrational motion, the contribution could rise to values equal to about R per degree of freedom. The way to proceed is not yet clear, but the examples of Table 6.3 indicate that the principal contributions to the heat capacity at constant volume of simple liquids have been recognized.

HARDNESS OF CRYSTALS

Although some of the electrons of a metal are "free" to conduct an electric current, they make only a small contribution to the heat capacity.

Metals are distinguished by their luster and their electrical conductivity. These characteristics lead directly to the idea of relatively free electrons moving throughout the crystal lattice. The study of metallic crystals becomes primarily the study of how these electrons can be described and how their behavior can be investigated.

The results presented show that the heat capacities of metallic crystals are of a magnitude and have a temperature dependence similar to that of simple ionic or covalent crystals. At first, this result should be surprising. The relatively free electrons of a metal might be expected to lead to an additional heat-capacity contribution of about 3(½R) if each atom of the metal contributes one electron that is not tightly bound to the atom. A closer look at the heat-capacity data for metals is therefore called for.

Data for the heat capacity of silver down to low temperatures are studied. Close inspection of these data shows that although at the higher temperatures they are in line with the Debye theory and to a lesser extent with the Einstein theory, they are not consistent with these theories at the lower temperatures. In fact, at temperatures low enough for the Debye law to reduce to the T^3 relation, the data can be fitted by an empirical equation of the form

$$C_v = \alpha T^3 + \gamma T.$$

The first term can be looked on as arising from the lattice vibrations, in the manner shown by Debye and dealt. The linear-temperature term must be attributed to the electronic contribution of the metallic crystal.

MEASUREMENT OF VAPOUR PRESSURE

Raoult measured the individual vapour pressures of a liquid and then the solution by the method. He introduced the liquid or the solution into Toricellian vacuum of a barometer tube and measured the depression of the mercury level. This method is neither practicable nor accurate as the lowering of vapour pressure is too small.

Manometric Method

The vapour pressure of a liquid or solution can be conveniently measured with the help of a manometer The bulb B is charged with the liquid or solution. The air in the connecting tube is then removed with a vacuum pump. When the stopcock is closed, the pressure inside is due only to the vapour evaporating from the solution or liquid. This method is generally used for aqueous solutions. The manometric liquid can be mercury or n-butyl phthalate which has low density and low volatility. Now let us learn about, gas saturation method for the measurement of lowering of vapour pressure.

Ostwald and Walker's Dynamic Method *(Gas Saturation Method)*

In this method the relative lowering of vapour pressure can be determined straightaway. The measurement of the individual vapour pressures of a solution and solvent is thus eliminated.

Procedure

It consists of two sets of bulbs :

(a) Set A containing the solution,

(b) Set B containing the solvent.

Each set is weighed separately. A slow stream of dry air is then drawn by suction pump through the two sets of bulbs. At the end of the operation, these sets are reweighed. From the loss of weight in each of the two sets, the lowering of vapour pressure is calculated. The temperature of the air, the solution and the solvent must be kept constant throughout.

Calculations

As the air bubbles through set A it is saturated up to the vapour pressure p_s of solution and then up to vapour pressure p of solvent in set B. Thus, the amount of solvent taken up in set A is proportional to p_s and the amount taken up in set B is proportional to $(p - p_s)$.

If w_1 and w_2, be the loss of weight in set A and B respectively,

$$w_1 \propto p_s \qquad ...(1)$$

$$w_2 \propto p - p_s \qquad ...(2)$$

Adding (1) and (2), we have

$$w_1 + w_2 \propto p_s + p - p_s$$

$$\propto p \qquad ...(3)$$

dividing (2) by (3) we get

$$\frac{p - p_s}{p} = \frac{w_2}{w_1 + w_2}$$

BOILING POINT ELEVATION

When a liquid is heated its vapour pressure rises and when it equals the atmospheric pressure the liquid boils.

If T_b is the boiling point of the solvent, and T is the boiling point of the solution the difference in the boiling points (ΔT) is called the elevation of boiling point.

$$T - T_b = \Delta T.$$

The vapour pressure curves are shown in the Figure 5.10.

For dilute solutions the curves BD and CE are parallel and straight lines approximately. Therefore, for similar triangles ACW and ABD we have

$$\frac{AB}{AC} = \frac{AD}{AE}$$

or

$$\frac{T_1}{T_2 - T_b} = \frac{p - p_1}{p - p_2}$$

Or this can be written as

$$\Delta T \propto p - p_s.$$

($p - p_1$) and $p - p_2$ are lowering of vapour pressure for solution 1 and solution 2.)

So, the elevation of boiling point, is directly proportional to the lowering of vapour pressure.

Determination of Molecular Mass from Elevation of Boiling Point

Since p is constant for the same solvent at a fixed temperature, from (1) we can write

$$\Delta T \propto \frac{p - p_s}{p}$$

But from Raoult's Law for dilute solutions,

$$\frac{p - p_s}{p} \propto \frac{wM}{Wm}$$

Since M (mol mass of solvent) is constant, from (3)

$$\frac{p - p_s}{p} \propto \frac{w}{Wm}$$

From (2) and (4)

$$\Delta T \propto \frac{w}{m} \times \frac{1}{W}$$

$$\Delta T = K_b \times \frac{w}{m} \times \frac{1}{W}$$

where K_b is a constant called *Boiling Point Constant* or *Ebbulioscopic Constant* or *Molal Elevation Constant.* If w/m = 1 and W = 1, K_b = AT. Thus,

Molal Elevation Constant may be defined as the boiling-point elevation produced when 1 mole of solute is dissolved in one kg (1000g) of the solvent.

If the mass of the solvent (W) is given in grams, it has to be converted into kilograms. Thus, the expression (5) assumes the form

$$\Delta T = K_b \times \frac{w}{m} \times \frac{1}{W/1000}$$

whence

$$m = \frac{1000 \times K_b \times w}{\Delta T \times W}$$

where ΔT = elevation of boiling point; K,, = molal elevation constant; w = mass of solute in grams. m = mol mass of solute; and W = mass of solvent in grams.

Sometimes the value of K_b is given in K per 0.1 kg (100 g). In that case, the expression becomes

$$m = \frac{100 \times K_b \times w}{\Delta T \times W}$$

The Value of K_b : The value of K_b can be determined by measurement of ΔT by taking a solute of known molecular mass (m) and substituting the values in expression (7).

Units of K_b *:* From equation, we have

$$K_b = \frac{\Delta T \times W/1000}{w/m}$$

$$= \frac{\Delta T \times \text{kg solvent}}{\text{mol solute}}$$

Thus the units of K_b are

$$\frac{^oCkg\ solvent}{mol\ solute}$$

The constant K_b, which is characteristic of a particular solvent used, can also be calculated from thermodynamically derived relationship

$$K_b = \frac{RT_b^2}{1000 \times L_v}$$

where R = gas constant;

T_b, = boiling point of solvent;

L_v = molar latent heat of vaporization.

Thus, for water R = 8.314 J mol^{-1}; T = 373 K; L_V = 2260 Jg^{-1}.

Therefore,

$$K_b = \frac{8.314 \times 373 \times 373}{1000 \times 2260} = 0.52\ L\ kg^{-1}$$

Molal Boiling-Point Constants

Solvent	K_b per kg (1000g)	K_b per 0.1 kg (100g)
Water	0.52	5.2
Propanone (acetone)	1.70	17.0
Ethoxyethane (ether)	2.16	21.6
Ethanoic acid (acetic acid)	3.07	30.7
Ethanol	1.75	11.5
Benzene	2.70	27.0
Trichloromethane (chloroform)	3.67	36.7

MEASUREMENT OF BOILING POINT ELEVATION

Several methods are available for the measurement of the elevation of boiling point. Now let us study about.

Landsberger-Warker Method

The apparatus is arranged as shown in the figure.

Pure solvent is placed in the graduated tube and vapour of the same solvent boiling in a separate flask is passed into it. The vapour causes the solvent in the tube to boil by its latent heat of condensation. When the solvent starts boiling and temperature becomes constant its boiling point is recorded. Now the supply of vapour is temporarily cut off and a weighed pellet of the solute is dropped into the solvent in the inner tube. The solvent vapour is again passed through until the boiling point of the solution is reached, and this is recorded.

The solvent vapour is then cut off thermometer and rosehead raised out of the solution and the volume of the solution read.

Molecular weight of the solution can be calculated as follows:

$$m = \frac{1000 \times K_b \times w}{\Delta T \times W}$$

where,

w = weight of the solute

W = weight of the solvent.

Now let us study about *Cottrell's Method.*

A method better than Landsberger-Walker method was devised by Cottrell (1910).

Apparatus

It consists of:

(i) A graduated boiling tube containing solvent or solution;

(ii) A reflux condenser which returns the vapourised solvent to the boiling tube;

(iii) A thermometer reading to 0.01 K, enclosed in a glass hood;

(iv) A small inverted funnel with a narrow stem which branches into three jets projecting at the thermometer bulb.

Beckmann Thermometer

It is a differential thermometer. It is designed to measure small changes in temperature and not the temperature itself. It has a large bulb at the bottom of a fine capillary tube. The scale is calibrated from 0 to 6 K and subdivided into 0.01 K.

The unique feature of this thermometer, however, is the small reservoir of mercury at the top. The amount of mercury in this reservoir can be decreased or increased by tapping the thermometer gently. In this way the thermometer is adjusted so that the level of mercury thread will rest at any desired point on the scale when the instrument is placed in the boiling (or freezing) solvent.

Procedure

Solvent is placed in the boiling tube with a porcelain piece lying in it. It is heated on a small flame (micro burner). As the solution starts boiling, solvent vapour arising from the porcelain piece pump the boiling liquid into the narrow stem. Thus, a mixture of solvent vapour and boiling liquid is continuously sprayed around the thermometer bulb. The temperature soon becomes constant and the boiling point of the pure solvent is recorded.

Now a weighed pellet of the solute is added to the solvent and the boiling point of the solution noted as the temperature becomes steady. Also, the volume of the solution in the boiling tube is noted. The difference of the boiling temperatures of the solvent and solute gives the elevation of boiling point. While calculating the molecular weight of solute the volume of solution is converted into mass by multiplying with density of solvent at its boiling point.

BECKMANN'S METHOD

The depression of freezing point can be measured correctly by Beckmann's method.

Apparatus

It consists of:

(i) A *freezing tube* with a side-arm to contain the solvent or solution, while the solute can be introduced through the side-arm;

(ii) An *outer larger tube* into which is fixed the freezing tube, the space in between providing an air jacket which ensures a slower and more uniform rate of cooling;

(iii) A *large jar* containing a freezing mixture *e.g.*, ice and salt, and having a stirrer.

Procedure

15 10 20 g of the solvent is taken in the freezing point tube and the apparatus set up so that the bulb of the thermometer is completely immersed in the solvent First determine the approximate freezing point of the solvent by directly cooling the freezing-point tube in the cooling bath. When this has been done, melt the solvent and place the freezing-point tube again in the freezing bath and allow the temperature to fall.

When it has come down to within about a degree of the approximate freezing point determined above, dry the tube and place it cautiously in the air jacket. Let the temperature fall slowly and when it has come down again to about 0.5° below the freezing point, stir vigorously. This will cause the solid to separate and the temperature will rise owing to the latent heat set free. Note the highest temperature reached and repeat the process to get concordant value of freezing point.

The freezing point of the solvent having been accurately determined, the solvent is remelted by removing the tube from the bath, and a weighed amount (0.1 – 0.2 g) of the solute is introduced through the side tube. Now the freezing point of the solution is determined in the same way as that of the solvent A further quantity of solute may then be added and another reading taken. Knowing the depression of the freezing point, the molecular weight of the solute can be determined by using the expression

$$m = \frac{1000 \times K_f \times w}{\Delta T \times W}$$

This method gives accurate results, if the following precautions are observed:

(i) The supercooling should not exceed 0.5°C.

(ii) The stirring should be uniform at the rate of about one movement per second.

(iii) The temperature of the cooling bath should not be 4° to 5° below the freezing point of the liquid.

In 1922, Rast improved Beckmann's Method. It was called Rast's Camphor Method. Rast used his method for, determination of molecular weights of solutes, which are soluble in Molten Camphor.

OSMOTIC PRESSURE

The Osmotic Pressure developed between a solvent and a solution depends only on the mole fractions of the components of the solution and properties of the solvent. The phenomenon of Osmosis depends on the existence of semi permeable membranes. Such membranes are of many types but they are all characterised by the fact that they allow one component of a solution to pass through and prevent the passage of another component. Cellophane and a number of plant or animal membranes, for example, are permeable to water but not to higher-molecular-mass compounds. Mention can also be made of the semipermeability of a palladium foil, which is permeable to hydrogen gas but not to nitrogen and other gases. With such a membrane, osmosis can be studied in the vapour phase.

An osmosis apparatus depends on the separation of a solution from its pure solvent by means of a membrane, permeable to the solvent but impermeable to the solute. The essential features of the system. When such an arrangement is made, there is a natural tendency for the solvent to flow from the pure-solvent chamber through the membrane into the solution chamber. This tendency can be opposed by applying pressure to the solution chamber. The excess pressure that must be applied to the solution to produce equilibrium is known as the osmotic pressure. It is through this quantity that the quantitative aspects of osmosis are studied.

The osmotic pressure developed between any dilute solution and its solvent will be shown to be a colligative property. It is therefore dependent on only the concentration of the solution and the properties of the solvent. It is important to recognize that the nature of the semipermeable membrane and the mechanism by which it allows solvent to pass through it but prevents the passage of solute is of no importance for the study of osmotic pressure as a colligative property. The thermodynamic basis of the osmotic pressure can now be shown. The solvent in the pure-solvent compartment will be subject to no pressure or temperature change or solute addition. Thus,

$$d(G\ A)_{pure\ solv} = 0$$

The solvent in the solution compartment, however, is subject to the addition of solute and a pressure change. The free energy changes as a result of the change in x_A, assuming ideal behaviour, according to an RT $d(\ln x_A)$ term. The free-energy change due to a pressure change dP is $\overline{V}$ dP, where V_A is the partial molal volume of A. Thus, for the solvent in the solution compartment

$$d\overline{G} = \overline{V}_A\, dP + RT\, d \ln x_A$$

For equilibrium to be maintained with the unaffected pure solvent of the solvent compartment, the net change in $\overline{G}_A$ must be zero. Thus

$$\overline{V}_A\, dP = RT\, d \ln x_A$$

Integration from $x_A = 1$, for which the pressure is P_{init} to x_A, for which the pressure is P_{final}, can be carried out if it is assumed that $\overline{V}_A$ is a constant equal to the molal volume V_A of the pure solvent. Then

$$V_A(P_{final} - P_{init}) = -RT \ln x_A$$

The excess pressure is the osmotic pressure II. For $x_A = 1 - x_B$ and $x_B \leq 1$, and $\ln x_A \approx -x_B$, equation gives the desired colligative-property relation

$$\Pi V = RTx_B$$

An interesting variation of this result is obtained by multiplying by n_A, the number of moles of A, which is approximately equal to $n_A + n_B$. This gives

$$n_A V_A \Pi = RTn_A x_B \approx n_B RT$$

If V is introduced to represent $n_A V_A$, which is approximately the volume of solution containing rig mol of solute, we can write

$$\Pi V = n_B RT$$

or

$$\Pi = MRT,$$

where M is the molarity of the solution. The similarity of the first of these expressions to the Ideal-Gas Law led Van't Hoff and others to some not very fruitful ideas that view the osmotic pressure as arising from a molecular-bombardment process. It is recognized here that equations are merely approximate forms obtained from the thermodynamic dilute-solution expression.

GIBBS DUHEM EQUATION

Partial molal properties of one component of a binary solution are related to those of the other component through the Gibbs-Duhem equation.

It is often not feasible to determine the partial molar free energies and the activities of solutes from measurements of the vapour pressure of the solute. For non-volatile solutes, for example, it is clearly necessary to have an alternative procedure, and here one that depends on the relation between solute and solvent properties as given by the *Gibbs-Duhem equation* is developed.

You will see that if the composition of a solution is changed, the changes in the partial molal properties of the components of the solution are interrelated. This general result lets us use changes in the partial molal properties of one component to deduce changes in this property of the other component.

The Gibbs-Duhem relation is developed in terms of free energies. The results apply to any thermodynamic property, as is made clear by the use of volumes to illustrate the relations.

The free energy of a mixture, or solution, at a particular temperature and pressure is a function of the number of moles of the components.

Changes in the free energy of the solution are related to the amount of each of the *i*th components by the total differential

$$dG = \sum_i \frac{\partial G}{\partial n_i} dn_i$$

$$= \sum_i \bar{G}_i \, dn_i$$

where the bar indicates the partial molal quantity.

The argument of Sec. 8-2, which led to equation, lets us write the free energy of the solution as

$$G = \sum \bar{G}_i \, dn_i$$

According to this description of the solution free energy, changes in this free energy are given by changes in the partial molal free energies or by changes in the number of moles of the components according to

$$dG = \sum_i \bar{G}_i \, dn_i + \sum_i n_i \, \bar{G}_i$$

It follows from equations and that

$$\sum_i n_i \, d\bar{G}_i = 0$$

We refer to this equation, or ones developed from it, as the *Gibbs-Duhem equation.* It applies to any of the thermodynamic properties of the components of a solution.

For a two-component solution, with components A and B, equation is

$$n_A d\bar{G}_A + n_B d\bar{G}_B = 0$$

or

$$n_A d\bar{G}_A = - n_B d\bar{G}_B$$

If the composition of the solution changes, the change in the partial molal free energy of component A is related to the changes in that property of component B.

Partial Molal Volumes

The Gibbs-Duhem relation has been derived in terms of partial molal free energies. It applies equally to other partial molal quantities. The easily visualized partial molal volumes can be used to illustrate the relation.

If we write equation for volumes instead of free energies and divide through by $n_A + n_B$, so that we can deal with mole fractions, we obtain

$$x_A d\bar{V}_A = -x_B d\bar{V}_B$$

For a change in the mole fraction of one of the components, A for example, this relation becomes

$$x_A \frac{d\bar{V}_A}{dx_A} = -x_B \frac{d\bar{V}_B}{dx_A}$$

This Gibbs-Duhem relation shows that the slopes of the partial-molal-volume curves for the two components of a binary solution are simply related.

Notice that at a mole fraction of 0.5 the slopes of the curves for the two components are equal and opposite. Furthermore, the slopes of curves at any mole fraction are of opposite sign. At mole fractions other than $x = 0.5$, the magnitude of the slope for the minor component is larger and that for the major component is smaller,

In the limit of zero mole fraction of a component, the slope of the partial-molal-quantity curve for that component must become infinite, unless the slope of the curve for the other component becomes zero. The partial molal volume of the solvent of dilute solutions. The solute partial molal volume remains finite even in the zero concentration limit.

PARTIAL MOLAL QUANTITIES

The properties of a component of a mixture are expressed as partial molal quantities. The volume of a solution formed by mixing 50 mL of water and 50 mL of ethanol is 95 mL. When 1 g of magnesium sulfate is added to 100 mL of water, the volume of the solution is less, by about 0.01 mL, than that of the water we started with. When acetone and chloroform are mixed, the solution feels appreciably warm. When methanol and carbon tetrachloride are mixed, the solution feels cool. Many real solutions are not ideal. They cannot be described by the simple addition of the properties of the components of the solution.

In preceding chapters you learned how to measure, interrelate, and use thermodynamic properties of substances. Now we need to extend the general procedures to mixtures of substances. We need methods that will let us deal with the enthalpy, entropy, and free energy of the components of a mixture of substances and of the mixture itself. Only then can we apply, for example, the relation between free energy and equilibria to reactions between the components of a solution. (In fact, all chemical equilibria involve a mixture of components. We assumed that the properties of components of a solution could be simply related to the properties of the pure substances. This is generally not the case.) Now we develop a procedure for using measurable properties of a solution to obtain quantities that we can attribute to the components of the solutions.

Partial Molal Volumes

We begin by seeing how we can assign parts of the volume of a binary solution to each of the two components.

The data that are generally available and can be used to obtain volume information are the densities of solutions. These are often reported for solutions with various amounts of the minor component, referred to as the *solute*, in some fixed amount of the major component, called the *solvent*. These density data can be used to calculate the volume of solutions with a given amount of solvent and various amounts of solute. Volumes for solutions of various *molalities m*, defined as the number of moles of solute per 1000 g of solvent, for aqueous solutions of ammonia.

We introduce the symbol n_A for the number of moles of solvent and n_B for the number of moles of solute. The slopes of curves give values of the partial derivative $\left(\frac{\partial V}{\partial n_B}\right)_{n_A}$ or, more completely, $\left(\frac{\partial V}{\partial n_B}\right)_{n_A T.P.}$

This type of derivatives with T, P and the amounts of the other, reagents held constant, are examples of partial molal quantities. They play a central role in the thermodynamic treatment of solutions.

The partial molal properties give the contribution of 1 mol of the material to the property of the solution being considered. The partial molal volume for example, is the change in volume of the solution, that accompanies the addition of 1 mol of component to a sample of the solution large enough that the concentration is not appreciably changed by this addition. We can therefore interpret the Partial Molal Volume as the volume due to, or assigned to, 1 mol of that component in the solution.

QUANTUM MECHANICS AND WAVE MECHANICS

One of the most exciting endeavours in our investigation of the world has been our attempt to understand the basic units of matter that make up the material world. For the chemist the basic units are the molecules and the atoms of which they are composed. Some of the long sought for answers to the questions of why and how atoms are held together into molecules can now be given. The description of the nature of chemical bonding that is achieved represents the culmination of one aspect of our efforts to unravel the secrets of matter.

The classical 'mechanical theory' of matter considered matter to be made of discrete particles elects, protons etc.), another theory called the 'Wave theory' was necessary to interpret the nature of radiations like X-rays and light. According to the wave theory, radiations as X-rays rays and light, consisted of continuous collection of waves travelling in space.

The wave nature of light, however, failed completely to explain the photoelectric effect *i.e.*, the emission of electrons from metal surfaces by the action of light. In their attempt to find a plausible explanation of radiations from heated bodies as also the photoelectric effect, Planck and Einstein (1905) proposed that energy radiations, including those of heat and light, are emitted discontinuously as little 'bursts', quanta, or photons. This view is directly opposed to the wave theory of light and it gives particle-like to waves. *According to it, light exhibits both a wave and a particle nature, under suitable conditions.* This theory which applies to all radiations, is often referred to as the *'Wave Mechanical Theory'*.

Bohr, undoubtedly, gave the first quantitative successful model of the atom. But now it has been superseded completely by the modern *Wave Mechanical Theory.* The new theory rejects the view that electrons move in closed orbits, as was visualised by Bohr. The Wave mechanical theory gave a major breakthrough by suggesting that the electron mechanical theory gave by its wave properties and probabilities.

With Planck's contention of light having wave and particle nature, the distinction between particles and waves became very hazy. In 1924 Louis de Broglie advanced a complimentary hypothesis for material particles. According to it, the dual character—the wave and particle—may not be confined to radiations alone but should be extended to matter as well. In other words, matter also possessed particle as well as wave character. This gave birth to the *'Wave mechanical theory of matter'*. This theory postulates A at electrons, protons and even atoms, when in motion, possessed wave properties and could also be associated with other characteristics of waves such as wavelength, wave-amplitude and frequency. *The new quantum mechanics, which takes into account the particulate and wave nature of matter, is termed the Wave mechanics.*

THE UNCERTAINTY PRINCIPLE

One of the most important consequences of the dual nature of matter is the uncertainty principle developed by Werner Heisenberg in 1927.

This principle is an important feature of wave mechanics and discusses the relationship between a pair of *conjugate properties* (those properties that are independent) of a substance. According to the uncertainty principle, it is impossible to know simultaneously both the conjugate properties accurately. For example, the position and momentum of a moving particle are interdependent and thus conjugate properties also. Both the position and the momentum of the particle at any instant cannot be determined with absolute exactness or certainty. If me momentum (or velocity) be measured very accurately, a measurement of the position of the particle correspondingly becomes less precise. On the other hand if position is determined with accuracy or precision, the momentum becomes less accurately known or uncertain. *Thus certainty of determination of one property Introduces uncertainty of determination of the other.* The uncertainty in measurement of position, Δx, and the uncertainty of determination of momentum, Δp (or Δmv), are related by Heisenberg's relationship as

or $$\Delta x \times m\Delta v \geq \frac{h}{2\pi}$$

where h is Planck's constant.

It may be pointed out here that there exists a clear difference between the behaviour of large objects like a stone and small particles such as electrons. *The uncertainty product is negligible in case of large objects.*

For a moving ball of iron weighing 500g, the uncertainty expression assumes the form

$$\Delta x \times m\Delta v \geq \frac{h}{2\pi}$$

or $$\Delta x \times \Delta v \geq \frac{h}{2\pi m}$$

$$\geq \frac{6.625 \times 10^{-27}}{2 \times 3.14 \times 500} \approx 5 \times 10^{-31} \text{ erg sec g}^{-1}$$

which is very small and thus negligible. Therefore for large objects, the uncertainty of measurements is practically nil

But for an electron of mass $m = 9.109 \times 10^{-28}$g, the product of the uncertainty of measurements, is quite large as

$$\Delta x \times \Delta v \geq \frac{h}{2\pi m}$$

$$\geq \frac{6.625 \times 10^{-27}}{2 \times 3.14 \times 9.109 \times 10^{-28}} \approx 0.3 \text{ erg sec g}^{-1} \text{(approx.)}$$

This value is large enough in comparison with the size of the electron and is thus in no way negligible. If position is known quite accurately *i.e.*, Δx is very small, the uncertainty regarding velocity Δv becomes immensely large and *vice versa*. It is therefore very clear that the uncertainty principle is only important in considering measurements of small particles comprising an atomic system.

CATHODE RAYS

In 1986 J.J. Thomson, studied about cathode rays through his discharge tube experiment.

The discharge tube consists of a glass tube with metal electrodes fused in the walls. Through a glass, side arm air can be drawn with a pump.

The electrodes are connected to a source of high voltage (10,000 Volts) and the air partially evacuated. The electric discharge passes between the electrodes and the residual gas in the tube begins to low. If virtually all the gas is evacuated from within the tube, the glow is replaced by faintly luminous 'rays' which produce fluorescence on the glass at the end far from the cathode. *The rays which proceed from the cathode and move away from it at right angles in straight lines, are called Cathode Rays.*

Properties of Cathode Rays

(1) They travel in straight lines away from the cathode and cast shadows of metallic objects placed in their path.

(2) Cathode rays cause mechanical motion of a small pin-wheel placed in their path. Thin hey possess kinetic energy and must be material particles.

(3) They produce fluorescence (a glow) when they strike the glass wall of the discharge tube.

(4) They heat up a metal foil to incandescence which they impinge upon.

(5) Cathode rays produce X-rays when they strike a metallic target

(6) Cathode rays are deflected by the electric as well as the magnetic field in a way indicating that they are streams of minute particles carrying negative charge.

By counterbalancing the effect of magnetic and electric field on cathode rays, Thomson was able to work out the ratio of the charge and mass (e/m) of the cathode particle. In SI units the value of e/m of cathode particles is -1.76×18^8 coulombs per gram. As a result of several experiments, Thomson showed that the value of e/m of the cathode particle was the same regardless of both the gas and the metal of which the cathode was made. This proved that the particles making up the cathode rays were all identical and were constituent parts of the various atoms. Dutch Physicist H.A. Lorentz named them *Electrons*.

Electrons are also obtained by the action of X-rays or ultraviolet light on metals and from heated filaments. These are also emitted as β-particles by radioactive substances. Thus it is concluded that *electrons are a universal constituent of all atoms.*

Positive Rays

In 1886 Eugen Goldstein used a discharge tube with a hole in the cathode He noted that while cathode rays were streaming away from the cathode, there were coloured rays produced simultaneously which passed through the perforated cathode and caused a glow on the wall opposite to the anode. Thomson studied these rays and showed that they consisted of particles carrying a positive charge. He called them *Positive rays.*

Properties of Positive Rays

(1) They travel in a straight line in a direction opposite to the cathode.

(2) They are deflected by electric as well as magnetic field in a way indicating that they are positively charged.

(3) The charge-to-mass ratio (e/m) of positive particles varies with the nature of the gas placed in the discharge tube.

(4) They possess mass many times the mass of an electron.

(5) They cause fluorescence in zinc sulphide.

When high-speed electrons (cathode rays) strike molecule of a gas placed in the discharge tube, they knock out one or more electrons from it. Thus a positive ion results

$$M + e^- \rightarrow M^+ + 2e^-.$$

These positive ions pass through the perforated cathode and appear as positive rays. When electric discharge is passed through the gas under high electric pressure, its molecules are dissociated into atoms and the positive atoms (ions) constitute the positive rays

From a study of the properties of positive rays, Thomson and Aston (1913) concluded that atom consists of at least two parts :

(a) the electrons; and

(b) a positive residue with which the mass of the atom is associated.

BOHR ATOMIC MODEL

The Bohr theory of the hydrogen atom, before the general development of quantum mechanics, was based on the quantization of angular momentum. The beginnings of our modern understanding of the electronic structure of atoms and molecules can be associated with the treatment of the hydrogen atom given by Niels Bohr in 1913. Bohr arbitrarily introduced the idea that the angular momentum of the electron, which he pictured as moving in a circular orbit about the nucleus, could have only certain angular-momentum values. We now know that a path or trajectory of an electron of an atom or molecule cannot be determined, and should not be part of a theoretical development. The *Bohr atom*. In spite of such shortcomings, introduces a number of features of value.

The following ideas, or postulates, provide the basis for the Bohr description of the hydrogen atom:

1. The electron moves about the nucleus in a circular orbit.
2. Only orbits in which the electron has an angular momentum that is an integral multiple of $h/(2\pi)$, or $\hbar$ are allowed.
3. The electron does not radiate energy when it is in an allowed orbit, but it can gain or lose energy by moving from one orbit to another.

On the basis of these ideas, the allowed radii and energies of a hydrogen atom can be deduced.

Neils Bohr, a brilliant Danish Physicist, pointed out that the old laws of physics just did not work in the submicroscopic world of the atom. He closely studied the behaviour of electrons, radiations and atomic spectra. Bohr proposed anew model of the atom based on the modem Quantum theory of energy. With his theoretical model he was able to explain as to why an orbiting electron did not collapse into the nucleus and how the atomic spectra were caused by the radiations emitted when electrons moved from one orbit to the other. Therefore to understand the Bohr theory of the atomic structure, it is first necessary to acquaint ourselves with the nature of electromagnetic radiations and the atomic spectra as also the Quantum theory of energy.

ATOMIC SPECTRA

When an element in the vapour is heated in a flame or discharge tube, the atoms are excited and emit light radiations of a characteristic colour. The colour of light produced indicates, the wave length of the radiation emitted.

For example, a Bunsen burner flame is coloured yellow by sodium salts, red by strontium and violet by potassium. In a discharge tube, neon glows orange-red, helium pink, and so on. If we examine the emitted light with a *Spectroscope* (a device in which a beam of light is passed through a prism and received on a photograph), the spectrum obtained on the photographic plate is found to consist of bright lines. Such a spectrum in which each line represents a specific wavelength of radiation emitted by the atoms is referred to as the Line spectrum or Atomic Emission spectrum of the element. An individual line of these spectra is called a Spectral line. When white light exposed of all visible wavelengths, is passed through the cool vapour of an element, certain wavelengths may be absorbed. These absorbed wavelengths are thus found missing in the transmitted light. The spectrum obtained in this way consists of a series of dark lines which is referred to as the *Atomic Absorption spectrum* or simply *Absorption spectrum*. The wavelengths of the dark lines are exactly the same as those of bright lines in the emission spectrum. The absorption spectrum of an element is the reverse of emission spectrum of the element.

The emission line spectrum of hydrogen can be obtained by passing electric discharge through the gas contained in a discharge tube at low pressure. The light radiation emitted is then examined with the help of a spectroscope. The bright lines recorded on the phonographic plate constitute the atomic spectrum of hydrogen.

In 1884 J.J. Balmer observed that there were four prominent coloured lines in the visible hydrogen spectrum:

(1) a *red line* with a wavelength of 6563 Å.

(2) a *blue-green line* with a wavelength 4861 Å.

(3) a *blue line* with a wavelength 4340 Å.

(4) a *violet line* with a wavelength 4102 Å.

The above series of four lines in the visible spectrum of hydrogen was named as the *Balmer Series*. By carefully studying the wavelengths of the observed lines, Balmer was able empirically to give an equation which related the wavelengths (K) of the observed lines. The *Balmer Equation* is

$$\frac{1}{\lambda} = R\left(\frac{1}{2^2} - \frac{1}{n^2}\right)$$

In the atomic spectrum of hydrogen five spectral series can be observed. They are as follows.

	Name	Region where located
(1)	Lyman Series	Ultraviolet
(2)	Balmer Series	Visible
(3)	Paschen Series	Infrared
(4)	Brackett Series	Infrared
(5)	Pfund Series	Infrared

PHOTOELECTRIC EFFECT

When a beam of light of sufficiently high frequency is allowed to strike a metal surface in vacuum, electrons are ejected from the metal surface. This phenomenon is known as Photoelectric effect and the ejected electrons Photoelectrons. For example, when ultraviolet light shines on Cs (or Li, Na, K, Rb) as in the apparatus the photoelectric effect occurs. With the help of this photoelectric apparatus the following observations can be made:

(1) An increase in the intensity of incident light does not increase the energy of the photoelectrons. It merely increases their rate of emission.

(2) The kinetic energy of the photoelectrons increases linearly with the frequency of the incident light. If the frequency is decreased below a certain critical value (Threshold frequency, V_0), no electrons are ejected at all.

The Classical Physics predicts that the kinetic energy of the photoelectrons should depend on the intensity of light and not on the frequency. Thus it fails to explain the above observations.

In 1905 Albert Einstein interpreted the photoelectric effect by application of the Quantum theory of light.

ZEEMAN EFFECT

In 1896 Zeeman discovered that spectral lines are split up into components when the source emitting lines is placed in a strong magnetic field. This phenomenon was called the *Zeeman effect* after the name of the discoverer.

It consists of electromagnets capable of producing strong magnetic field with pole pieces through which holes have been made length-wise. Let a discharge tube or sodium vapour lamp emitting radiations be placed between the pole pieces. When the spectral lines are viewed axially through the hole in the pole pieces *i.e.*, parallel to the magnetic field, the line is found to split up into two components, one having shorter wavelength (higher frequency) and the other having higher wave-length (shorter frequency) than that of the original spectral line, which is no longer observable. The two lines are symmetrically situated around the position of the original line and the change in wavelength is, termed the Zeeman shift (denoted as $d\lambda$). When viewed in a direction perpendicular to the applied field the lines split up into three, the central one having the same wavelength and frequency as that of the original line and the other two occupying the same position as observed earlier.

In order to explain Zeeman effect, let us consider motion of an electron in a particular orbit corresponding to its permitted angular momentum. The motion of the electron in an orbit is equivalent to a current in a loop of wire. If a current carrying loop of wire be placed in a magnetic field, it experiences a torque, and energy of the system depends upon the orientation of the loop with respect to magnetic field. The correct values of the energies are obtained if the components of the angular momentum of the electron along the direction of the magnetic field are restricted to the value

$$= m \times \frac{h}{2\pi}$$

where $m = 0, \pm 1, \pm 2,...$ and so on. Corresponding to these values of m, a given line splits into as many lines. Hence for each frequency of a radiation emitted by the atom in the absence of magnetic field, there are several possible frequencies in the presence of it This is, in fact, the cause of Zeeman Effect.

The shift in the frequency $d\lambda$ for each of the component lines is given by Lorentz's theoretically derived equation as

$$\frac{e}{m} = \pm \frac{4\pi c d\lambda}{h\lambda^2}$$

where H is the strength of magnetic field, e the electronic charge, m the mass of electron, c the velocity of light and λ the wavelength of the original line in the absence of magnetic field. The equation can also be written as

$$\text{Zeeman shift } d\lambda = \pm \frac{He\lambda^2}{4\pi mc}$$

The validity of the above equation can be tested experimentally by observing the Zeeman shift dλ for a given light source of known λ for a magnetic field of known strength H and calculating the value of e/m from the above equation. Lorentz found that the e/m of the electrons found by this method comes out to be the same as by any other method.

SELF CONSISTENT FIELD METHOD

The properties of many-electron atoms can be calculated by the "self-consistent" field method.

Just as an exact wave function could not be obtained for the two-electron system of the helium atom, so also exact solution functions cannot be found for atoms with more than two electrons. Furthermore, the hamiltonians for such atoms contain many troublesome r_{12}-type terms. In response to these difficulties, D.R. Hartree and, later, V.A. Fock developed a variation treatment described as a *self-consistent field (SCF) method.*

An approximate hamiltonian is set up so that specific coordinates of the r_{12} type which describe the distance between electrons are absent. Each electron is treated as moving in a potential field due to the nucleus and the averaged charge distribution due to all the other electrons of the atom.

This distribution is initially not known but trial functions can be used to calculate the probability function for each electron. The electron density corresponding to the trial function is then calculated.

This electron density contributes, along with the nuclear charge, a potential term that can be used to construct a hamiltonian for any one selected electron. Now the variation method can be used to improve the wave function for this single selected electron. This improved trial function can be used to calculate an improved charge-density contribution of this electron.

Next a second electron is selected, and its trial function is similarly improved. As this process is repeated, the improved wave functions and electron densities are fed back into the calculation. After enough cycles, the calculated energies and the trial functions do not change significantly, and we say that we have arrived at a self-consistent field.

The method fails to take into account the instantaneous electron-electron interactions. The positions and motions of the electrons are not "correlated" with each other. The determination of the *correlation energy* is an extremely difficult step still to be accomplished. Nevertheless, the SCF results are valuable, if approximate, guides to the orbitals and energies of many-electron atoms.

The orbitals that accommodate the outer, or valence, electrons in each periodic table row are now clearly shown.

ELECTRON CONFIGURATION

We have seen above that to define completely the state of an atom it is obligatory to refer to all the four quantum numbers (n, l, m and s) of every electron in it. Since a simultaneous representation of all quantum numbers of each electron in a single symbolic notation seems quite difficult, it is customary to take into account the first two quantum numbers only while the other two can be interred indirectly. The general symbolic notation employed for the purpose is nl^a where the numerical value of n = 1, 2, 3 etc., represents the principal quantum number, the letter designate of l (v for l = 0, p for l = 1 and so on) stands for the orbital and the superscript a gives the number of electrons in the orbital. Thus, $3s^2$ indicates that two electrons are present in the first subshell s (l = 0) of the third shell (n = 3). For instance, the distribution of seven electrons (of N atom) may be schematically represented as $1s^2$; $2s^2$, $2p^3$ or more elaborately as $1s^2$; $2s^2$, $2p_x^1$, $2p_y^1$, $2p_z^1$. By using the various designates of orbitals at a sublevel such as $2p_x$, $2p_y$ etc., the third quantum number *m* is also indicated (*e.g.*, $2p_x$ for m = +1, $2p_y$ for m = 0 and $2p_z$ for m = −1). Spin quantum numbers are indirectly inferred. Whenever there are two electrons in an orbital, one of these has +1/2 and the other −1/2 as their spin quantum number.

It is a common practice to denote an orbital by a horizontal line or a circle or square and an electron by an arrow over it. The direction of the arrow indicates the spin, an upward arrow representing a clockwise spin while the downward arrow stands for the anticlockwise direction of spin. When there are more than one orbitals in a subshell (*degenerate orbitals*), they are shown by an equivalent number of horizontal lines at the same energy level. Now let us study about the electron configuration of some elements.

HYDROGEN AND HELIUM

These have one and two electrons respectively which are accommodated in Is orbital while others remain vacant. The lone electron of hydrogen is filled in Is orbital and for helium the second electron would also go in 1s orbital, since it could accommodate another electron with opposite spin.

Lithium and Beryllium

These elements have three and four electrons respectively. The third electrons respectively. The third electron of L: enters in the 2s orbital and the fourth electron of Be also enters in the same orbital, but has an opposite direction of spin.

Boron and Carbon

These atoms have five and six electrons respectively. Is and 2s orbitals being completely filled with four electrons, the fifth electron of boron would go in one of the 2p orbitals say $2p_x$. The sixth electron in carbon would prefer to be accommodated in another vacant 2p orbital say ($2p_y$) rather than going to $2p_z$ orbital (Hund's rule). The two unpaired electrons shall have similar spins as indicated.

ELECTRONEGATIVITY

It is evident that an atom of high electronegativity will attract the shared electron pair away from one of lower electronegativity. Thus the former atom will acquire a partial negative charge while the other atom will get a partial positive charge.

The Pauling electronegativity values of some elements

Li	Be	B	C	N	0	F
1.0	1.5	2.0	2.5	3.0	3.5	4.0
Na	Mg	Al	Si	P	S	Cl
0.9	1.2	1.5	1.8	2.1	2.5	3.0
K	Ca	–	Ge	As	Se	Br
0.8	1.0		1.8	2.0	2.4	2.8

Rb	Sr	–	Sn	Sb	Te	I
0.8	1.0		1.8	1.9	2.1	2.5
Cs	Ba					
0.7	0.9					

Using measured values of bond energies, Pauling devised a set of electronegativity values. He allotted a value of 4 to the most electronegative atom, namely fluorine, and assigned values to the atoms of other elements.

Trends in Electronegativities

The variations in electronegativities of elements in the Periodic table are similar to those of ionisation energies and electron affinities.

(1) Increase Across a Period

The values of electronegativities increase as we pass from left to right in a Period. Thus, for Period 2 we have

Li	Be	B	C	N	0	F
1.0	1.5	2.0	2.5	3.0	3.5	4.0

This is so because the attraction of bonding electrons by an atom increases with increase of nuclear charge (At. No.) and decrease of atomic radius. Both these factors operate as we move to the right in a Period.

(2) Decrease Down a Group

The electronegativities of elements decrease from top to bottom in a Group. Thus, for Group VII we have

F	Cl	Br	I
4.0	3.0	2.8	2.5

The decrease trend is explained by more shielding electrons and larger atomic radius as we travel down a group.

Importance of Electronegativity

The electronegativities of elements are widely used throughout the study of Chemistry. Their usefulness will be discussed at appropriate places. The important applications of electronegativities are listed below.

(1) *In predicting the polarity of a particular bond :* The polarity of a bond, in turn, shows the way how the bond would break when attacked by an organic reagent.

(2) *In predicting the degree of ionic character of a covalent bond.*

Electron Affinity

A neutral atom can accept an electron to form a negative ion. In this process, in general, energy is released.

Electron affinity (EA) of an element is the amount of energy released when an electron is added to a gaseous atom to form an anion.

$$X_{(g)} + e \rightarrow X_{(g)} + EA$$

The energy involved in the addition of the first electron is called *first-electron affinity;* the energy involved in the addition of a second electron is called *second-electron affinity;* and so on. Thus,

$$X + e^{-} \rightarrow X^{-} + EA_1$$

$$X^{-} + e^{-} \rightarrow X^{2} + EA_2$$

The electron affinity of an element measures the ease with which it forms an anion in the gas phase.

Electron affinities are difficult to measure and accurate values are not known for all elements, They are expressed in kJ mol^{-1}.

ORIENTATION QUANTUM NUMBER

This quantum number has been proposed to account for the splitting up of spectral lines (Zeeman Effect). An application of a strong magnetic field to an atom reveals that electrons with the same values of principal quantum number 'n' and of azimuthal quantum number 'l', may still differ in their behaviour.

They must, therefore, be differentiated by introducing a new quantum number, the magnetic quantum number m. This is also called *Orientation Quantum Number* because it gives the orientation or distribution of the electron cloud. For each value of the azimuthal quantum number 'l', the magnetic quantum number m, may assume all the integral values between $+l$ to $-l$ through zero *i.e.,* $+l$, $(+l-1)$,...0...,$(-l+1)$, $-l$. Therefore for each value of l there will be (21 + 1) values of m_l. Thus when $l = 0$, m = 0 and no other value. This means that *for each value of principal quantum number 'n,' there is only one orientation for $l = 0$ (s orbital) or there is only one s orbital.* For s orbital, there being only one orientation, it must be spherically symmetrical about the nucleus. There is only one spherically symmetrical orbital for each value of n whose radius depends upon the value of n.

For $l = 1$ (p orbital), the magnetic quantum number m will have *three* values : +1, 0 and −1; so there are *three* orientations for *p* orbitals. These *three* types of *p* orbitals differ only in the value of magnetic quantum number and are designated as p_x, p_y, p_z depending upon the axis of orientation. The subscripts x, y and *z* refer to the co-ordinate axes. In the absence of a magnetic field, these three p orbitals are equivalent in energy and are said to be *three-fold degenerate* or *triply degenerate*. In presence of an external magnetic field the relative energies of the three p orbitals vary depending upon their orientation or magnetic quantum number. This probably accounts for the existence of more spectral lines under the influence of an external magnetic field. The p orbitals are of dumb-bell shape consisting of two lobes. The two lobes of a p oibital extend outwards and away from the nucleus along the axial line. Thus the two lobes of a p orbital may be separated by a plane that contains the nucleus and is perpendicular to the corresponding axis. Such plane is called a *nodal plane. There is no likelihood of finding the electron on this plane.* Foe a p_x orbital, the yz plane is the nodal plane. The shapes and orientations of the p orbitals.

For $l = 2$ (d orbital), the magnetic quantum numbers are *five* (2 × 2 + 1); + 2, + 1, 0, −1, −2. Thus there are five possible orientations for d orbitals which are equivalent in energy so long as the atom is not under the influence of a magnetic field and are said to be *five-fold degenerate*. The five d orbitals are designated as $d_{xy}, d_{yz}, d_{zx}, d_{x^2-y^2}, d_{z^2}$. These orbitals nave complex geometrical shapes as compared to p orbitals. The conventional boundary surfaces or shapes of five d orbitals. The shape of the d_z orbitals is different from others.

MOLECULAR TREATMENT

Many approaches, both theoretical and experimental, to the study of the behaviour of chemical systems have been developed and applied. In these studies a distinction has generally been made between a molecular treatment and a macroscopic one. An important and very interesting class of systems occurs which is, in a way, intermediate between these extremes. These systems often consist of or contain molecules so large that they can be treated either as large molecules or as small macroscopic particles. Most particles of current interest in this size range are found to be single molecules, and the term *macromolecule* is convenient.

Many of these systems, as will be seen, have great *biological importance*. Some of this area of study is often included in biochemistry, but the term molecular biology also seems appropriate. In the short study of macromolecules that can be presented here, it is desirable to discuss synthetic and naturally occurring macromolecules side by side. Although the areas of plastics and biological materials may seem little related, the physicochemical study of the basic chemical units of these areas has very much in common.

Many studies of macromolecules make use of solutions of the macromolecular material. And many of these studies depend on dynamic features of the particles in the solutions. Here, following introductory studies that treat the molecular mass and the shape of macromolecules in solution, we deal with the dynamic aspects of diffusion and the movement of macromolecules under the influence of centrifugal and electrical forces.

Colloids, macromolecules, and micelles lie between the categories of ordinary particles and ordinary molecules.

Colloids : The existence of particles in the size range dealt with here was suggested b the early observation made by botanist Robert Brown of the random motion of pollen grains as seen under a microscope. It was later recognized that these particles, though large enough to be seen, were small enough to reveal the effects of random molecular bombardment, the so-called *brownian motion*. By the end of the nineteenth century, study of small-particle systems, called *colloids,* became an important branch of physical chemistry.

The unique behaviour of colloids is now recognized to be exhibited by particles in the size range of about 10 to 1000 nm. One of the most commonly recognized features of such systems is that they scatter light, as is observed when a bean of sunlight passes through dusty air or through thin skimmed milk.

Furthermore, the particles of a colloidal system do not settle out and (as chemists experience with some silver chloride precipitates) tend to pass through ordinary filter paper.

Macromolecules : A close look at the chemical world shows that there are single molecules that are large enough for individual molecules to have colloidal dimensions. In view of present interest, these *macromolecules* can be classed either as synthetic polymers or as naturally occurring macromolecules. Most-interest in the natural materials is now centred on proteins and nucleic acids, but natural macromolecules also include the *polysaccharides* and the *polyisoprenes*, the latter being the molecules of natural rubber.

Micelles: The turbidity exhibited by soap or detergent solution is the best known indication of micelle formation. Most soaps and detergents have a long hydrocarbon "tail" and a polar 'head',. The molecules of soap and detergents are very small compared with colloidal dimensions. The particles causing the turbidity are groups or micelles of these molecules.

When the molecules, or particles, have a range of masses, the type of average that is used in an average mass must be specified.

Much of our information about the size and general shape of macromolecules has been deduced from various properties of solutions containing these molecules. In this section some methods used to understand the behaviour of solutions of macromolecules are investigated, and in the process of such studies a number of properties of the macromolecules themselves are discovered. It is first necessary, however, to discuss the meaning, and measurement, of molecular mass when it is applied to a polymeric material.

Polymerization reactions, both synthetic and natural, can lead to high-molecular-mass compounds. The reaction chain, however, is broken by some termination process that usually occurs in a random manner with respect to the size to which the polymer has already grown.

It follows that polymers have a range of molecular masses and that any data for the size or mass of the molecules of a polymer represent some sort of average value. It will be seen that attempts to deduce molecular masses of polymers lead to *number-average* and *mass-average* molecular masses.

OSMOSIS AND DIFFUSION

Just as a gas can diffuse into vacant space or another gas, a solute can diffuse from a solution into the pure solvent. If you pour a saturated aqueous solution of potassium permanganate with the help of a thistle funnel into a beaker containing water, it forms a separate layer at the bottom. After some time, you will see the permanganate actually diffusing up into water. This continues until a homogeneous solution is obtained.

The diffusion of solute into solvent is, in fact, a bilateral process. It consists of:

(1) the solute molecules moving up into solvent; and

(2) the solvent molecules moving down into solution. This intermingling of solute and solvent molecules goes on, so that ultimately a solution of uniform concentration results.

It is the tendency to equalise concentration in all parts of the solution which is responsible for the diffusion of the solute.

Thus diffusion of solute will also take place when two solutions of unequal concentrations are in contact. Solvent molecules will pass from the dilute to the concentrated solution and solute molecules will pass from the concentrated to the dilute solution until equality of concentration is achieved.

OSMOTIC PRESSURE

Now let us learn about osmotic pressure. A porous pot with cupric ferrocyanide membrane deposited in its walls is fitted with a rubber stopper having along glass tube. It is filled with concentrated aqueous sugar solution and immersed in distilled water. The osmosis of water through the membrane from water to the sugar solution takes place. As a result, the solution level in the long tube rises over a period of time. After a few days the level attains a definite maximum value. This marks the stage when the hydrostatic pressure set up due to the column of sugar solution counterbalances the flow of pure water (or osmosis) into the solution. *The hydrostatic pressure built up on the solution which just stops the osmosis of pure solvent into the solution through a semipermeable membrane, is called Osmotic Pressure.*

Osmosis can be counteracted not only by hydrostatic pressure but also by application of external pressure on the solution. The external pressure may be adjusted so as to prevent the osmosis of pure water into solution. This provides another definition of osmotic pressure.

Osmotic pressure may be defined as the external pressure applied to the solution in order to stop the osmosis of solvent into solution separated by a semipermeable membrane.

The external pressure on the solution is applied with the help of the piston and the progress of osmosis is shown by the movement of the liquid in the flow indicator. The pressure required just to arrest the movement of the liquid level in the (low indicator, is equal to the osmotic pressure of the solution.

Determination of Osmotic Pressure

The osmotic pressure of a given solution can be determined experimentally by the methods detailed below. The apparatus used for the purpose is often referred to as *osmometer*.

(1) Pfetter's Method

The apparatus used by Pfeffer (1877) for determination of osmotic pressure. It consists of a porous pot with copper ferrocyanide membrane deposited in its walls and having a glass top cemented to it. The glass top has an open tube and a closed mercury manometer in the side. The apparatus is filled with a solution under examination through the tube which is then sealed off The pot is now placed in pure water maintained at constant temperature. The water passes across the membrane into the solution and develops a pressure on the manometer. The highest pressure registered by the manometer gives the osmotic pressure of the solution.

The Pfeffer's Method Suffers from Two Disadvantages

(a) It is slow and it takes a few days before the highest pressure is reached.

(b) It cannot be used for measuring high osmotic pressures as the ferrocyanide membrane being weak ruptures.

OSMOTIC PRESSURE DETERMINATIONS OF MOLECULAR MASSES

A measurement of any of the colligative properties of a solution of macromolecules leads to a value for the number of solute molecules in a given amount of solvent. There it was pointed out that for the solutions of low molality, which are always obtained with macromolecules, the only colligative property that is conveniently measured is the osmotic pressure. Such measurements are one of the most important means of molecular-mass determinations. It should be evident from the nature of colligative properties, *i.e.*, their dependence only on the number and not on the nature of the solute molecule, that a number-average molecular mass is obtained.

The high concentration in terms of mass of solute per weight of solvent, even for low molalities, means that solute interactions will occur and nonideal behaviour will result. Therefore, it is almost always necessary to extrapolate the measurements to infinite dilution. Sensitive osmotic-pressure instruments now allow measurements to be made on very dilute solutions, and molar masses even up to 500,000 g can be obtained.

The expressions obtained for the osmotic pressure have been written in terms of mole fraction or molality. To study the osmotic pressure of a compound of unknown molecular mass, it is convenient to start with

$$\Pi = \frac{(n_B)RT}{n_A V} = \frac{(\text{mass of B})/(\text{mol mass of B})RT}{\text{volume of solution}}$$

Introduction of the concentration c as mass per unit volume lets the factor (mass of B)/(volume of solution) be represented by c. This allows the above expression to be written as

$$\frac{\Pi}{c} = \frac{RT}{\text{mol mass of solute}}$$

If the osmotic pressure is given in bar and c in grams per liter, the gas constant R must be entered as 0.08314 L bar K^{-1} mol^{-1}.

The can be expected to be valid only at infinite dilution. This follows, in addition to the approximations introduced in its derivation, from the fact that large-molecular-mass compounds tend to interact with each other at the concentrations at which the measurements are made. The procedure that must be used, therefore, is to measure Π/c as a function of c and to extrapolate these results to infinite dilution. The intercept of Π/c at zero concentration can then be taken as the value of RT/(mol mass of solute). From this value a valid molecular mass for the solute in solution is obtained.

OSMOSIS THEORIES

Some of the more important theories of osmosis can be explained as follows.

The Molecular Sieve Theory

According to this theory, the membrane contains lots of fine pores and acts as a sort of 'molecular sieve'. Smaller solvent molecules can pass through the pores but the larger molecules cannot Solvent molecules flow from a region of higher solute concentration to one of lower concentration across such a membrane. But since some membranes can act as sieves even though the solute molecules are smaller than the solvent molecules, this theory remains in doubt. Recently it has been shown that the pores or capillaries between the protein molecules constituting an animal membrane are lined with polar groups ($-COO^-$, $-NH_3^+$, $-S^{2-}$, etc.). Therefore, the membrane acts not simply as a sieve but also regulates the passage of solute molecules by electrostatic or 'chemical interactions'. In this way even solute molecules smaller than solvent molecules can be held back by the membrane.

Membrane Solution Theory

Membrane proteins bearing functional groups such as –COOH, –OH, –NH_2, etc., dissolve water molecules by hydrogen bonding or chemical interaction. Thus, membrane dissolves water from the pure water (solvent) forming what may be called 'membrane solution'. The dissolved water flows into the solution across the membrane in a bid to equalise concentrations. In this way water molecules pass through the membrane while solute molecules being insoluble in the membrane do not

Vapour Pressure Theory

It suggests that a semipermeable membrane has many fine holes or capillaries. The walls of these capillaries are not wetted by water (solvent) or solution.

Thus neither solution nor water can enter the capillaries. Therefore each capillary will have in it solution at one end and water at the other, separated by a small gap. *Since the vapour pressure of a solution is lower than that of the pure solvent, the diffusion of vapour will occur across the gap from water side to solution side.* This will result in the transfer of water into the solution. The vapour theory offers a satisfactory explanation of the mechanism of osmosis in most cases.

MEMBRANE BOMBARDMENT THEORY

This theory suggests that osmosis results from an unequal bombardment pressure caused by solvent molecules on the two sides of the semipermeable membrane. On one side we have only solvent molecules while on the other side there are solute molecules occupying some of the surface area. Thus there are fewer bombardments per unit area of surface on the solution side than on the solvent side. Hence the solvent molecules will diffuse more slowly through the membrane on the solution side than on the solvent side. The net result causes a flow of the solvent from the pure solvent to the solution across the membrane.

Reverse Osmosis

When a solution is separated from pure water by a semipermeable membrane, osmosis of water occurs from water to solution. This osmosis can be stopped by applying pressure equal to osmotic pressure, on the solution. If pressure greater than osmotic pressure is applied, osmosis is made to proceed in the reverse direction to ordinary osmosis *i.e.,* from solution to water.

The osmosis taking place from solution to pure water by application of pressure greater than osmotic pressure, on the solution, is termed Reverse Osmosis.

VAN'T HOFF THEORY

Van't Hoff noted the striking resemblance between the behaviour of dilute solutions and gases. Dilute solutions obeyed laws analogous to the gas laws. To explain it van't Hoff visualised that gases consist of molecules moving in vacant space (or vacuum), while in solutions the solute particles are moving in the solvent.

The pure solvent flows into the solution by osmosis across the semipermeable membrane. The solute molecules striking the membrane cause osmotic pressure and the sliding membrane is moved towards the solvent chamber. In case of a gas the gas molecules strike the piston and produce pressure that pushes it towards the empty chamber. Here, it is the vacuum which moves into the gas. This demonstrates clearly that there is close similarity between a gas and a dilute solution.

Thinking on these lines, van't Hoff propounded his theory of dilute solution. The *Van't Hoff Theory of Dilute Solutions* states that: a substance in solution behaves exactly like a gas and the osmotic pressure of a dilute solution is equal to the pressure which the solute .would exert if it were a gas at the same temperature occupying the same volume as the solution.

Suppose a dilute sugar solution to be contained in an exactly full, covered beaker. Then suppose the water to vanish, leaving the sugar molecules suspended in otherwise empty space. The sugar would then function as a gas, exerting pressure equal to its former osmotic pressure.

According to the van't Hoff theory of dilute solutions, all laws or relationships obeyed by gases would be applicable to dilute solutions.

From Van't Hoff theory it follows that just as 1 mole of a gas occupying 22.4 litres at 0°C exerts 1 atmosphere pressure, so 1 mole of any solute dissolved in 22.4 litres would exert 1 atmosphere osmotic pressure.

Van't hoff theory holds only for dilute solutions and if there is no dissociation or association of the solute molecules.

SEDIMENTATION COEFFICIENT

The sedimentation coefficient expresses the rate of movement in an ultracentrifuge study, and this quantity can be used, along with the diffusion coefficient, to estimate the macromolecular mass.

Macro molecules in solution can be made to alter their distribution in space by subjecting them to forces other than concentration gradients. It the macromolecule differs in density from the solvent, the simplest demonstration of this fact consists of allowing a solution to stand so that the force of gravity acts. A greater and more easily observed effect can be produced by using an ultracentrifuge. A sample of the macromolecule solution is rotated at a very high speed, in the neighbourhood of 10,000 to 80,000 r/min^{-1}.

Two essentially different types of ultracentrifuge experiments can be performed to determine the behaviour of solutions of macromolecules. In one, the rate of movement of the solute during the centrifugation is observed. In the other, the sample is centrifuged until an equilibrium distribution is obtained.

SEDIMENTATION VELOCITY

The first method starts with a well-defined boundary, or layer, of solution near the center of rotation and follows the movement of this layer toward the outside of the cell as a function of time. Such a method is termed a sedimentation-velocity experiment.

A particle of mass m at a distance x from the center of rotation experiences a force given by

$$f_{centrif} = m'x\omega^2$$

where ω is the angular velocity in radians per second and m' is the effective mass of the solute particle, *i.e.*, the actual mass corrected for the buoyancy effect of the solvent.

To express this buoyancy effect, we first recognize that v, the specific volume of the solute, is the volume of 1 g of solute. The volume of mg of solute is mv, and the mass of this volume of solvent is $m\upsilon\rho$, where ρ is the solvent density. Thus, the effective mass m' of the solute is $m - m\upsilon\rho = m(1 - \upsilon\rho)$. We now can rewrite

$$f_{centrif} = m(1 - \upsilon\rho)x\omega^2$$

This force is balanced for some constant-drift velocity dx/dt by a frictional force given by Stokes' law, as

$$f_{fric} = 6\pi r\eta \frac{dx}{dt}$$

Equating these two force expressions leads us to the constant-drift velocity. A rearrangement of the equality

$$m(1 - \upsilon\rho)x\omega^2 = 6\pi\eta r \frac{dx}{dt}$$

that collects the dynamic variables gives

$$\frac{dx/dt}{x\omega^2} = \frac{m(1 - \upsilon\rho)}{6\pi r\eta}$$

The collection of dynamic terms on the left side of describes the results of sedimentation-velocity experiments. This collection $(dx/dt)/x\omega^2$ can be looked on as the velocity with which the solute moves per unit centrifugal force, The *sedimentation coefficient* S is introduced as

$$S = \frac{dx/dt}{x\omega^2}$$

The experimental results can therefore be tabulated as values of S. The value of S for many macromolecules is of the order of 10^{-13}s. A convenient unit having the value 10^{-13} s has therefore been introduced, called a *svedberg*, in honor of T. Svedberg, who did much of the early work with the ultracentrifuge.

ELECTROPHORESIS

Many of the macromolecules of interest *e.g.*, proteins exist in solutions as charged species. It follows that movement of such particles can be produced by the application of an electric field. Such a process, known as electrophoresis, is the most important example of the more general electric-field motion effects described as electrokinetic. Studies of the migration of charged solute particles under the influence of an applied electrical potential can be carried out in a *moving-boundary* electrophoresis apparatus.

Depending on the charge carried by the particles, there will be a general movement of these species toward one electrode or the other. As a result, one of the original boundaries will move up, and the other will move down. Measurement of the movement of the boundary with time allows the mobility of the macromolecules to be deduced, at the pH of the buffer and at the temperature of the system. The mobility, defined as the rate of migration per unit voltage gradient, is given in units of m^2 V^{-1} s^{-1}.

The mobility of a particle in an electrophoresis experiment is the velocity reached when the frictional drag just balances the electric force. The frictional drag is given approximately for spherical particles by Stokes' law. The electric force is the charge eZ on the particle times the electrical potential gradient. The mobility u, which corresponds to a potential gradient of 1 V m^{-1}, can then be estimated from the relation

$$6\pi r\eta u = eZ \times (1\ V\ m^{-1})$$

or

$$u = \frac{eZ(1\,V\,m^{-1})}{6\pi r\eta}$$

ISOIONIC POINTS

The observed magnitude of protein mobilities and their dependence on pH are illustrated by the mobility-pH curve for β-lactoglobulin. The magnitude of the mobility at various pH values can be compared with typical mobilities of simple ions. By comparison, protein molecules, particularly in the multicharged forms that exist at high or low pH, are not as resistant to movement as might be expected.

The existence of a pH for which the mobility is zero. This pH is known as the *isoelectric point*. A closely related quantity is the isoionic point, defined as the pH at which the macromolecule carries no net charge as a result of proton gains or losses in acid-base reactions. The isoelectric and the *isoionic points* can differ because of the net charge that might result from the ion atmosphere forming about the protein molecule when it is in a buffer solution. In practice, the pH of the isoelectric and that calculated for the isoionic point are usually little different.

3

Free Energy

INTRODUCTION

Free energy is a property that provides a convenient measure of the driving force of a reaction.

The entropy change that must be considered if the direction of a chemical reaction is to be deduced is that of the universe of the reaction. This entropy change is the sum of that occurring in the system and that occurring in the thermal surroundings. Both contributions can be calculated from changes in the properties of the system.

Consider a chemical system in which a reaction occurs at constant temperature and constant pressure. The entropy change in the system is represented by ΔS. The entropy change in the thermal surroundings is calculated from the energy change in these surroundings. If we are dealing with an "ordinary" chemical reaction in which no mechanical energy, other than P dV type of energy is

involved, then the energy change of the thermal surroundings, is equal to $-\Delta H$. The entropy change in the thermal surroundings $\Delta U(\text{therm})/T$ is $\Delta H/T$.

The entropy change of the universe of the reaction system is given by

$$\Delta S_{univ} = \Delta S + \Delta S_{therm} = \Delta S - \frac{\Delta H}{T}$$

Notice that ΔS_{univ} can be calculated from changes in the properties of the system.

Free-Energy Property

We can modify above equation to arrive at a new, convenient system property. First, to convert the expression to one with the more familiar units of energy, we multiply by T, the constant temperature of the reaction. This gives

$$T\,\Delta S_{univ} = T\Delta S - \Delta H$$

Then, recognizing that $T\,\Delta S_{univ}$ is equated to an expression involving properties of the system, we introduce a new system property term, the *free energy*, with symbol G. Changes in this free energy, for a constant-temperature process, are given by

$$-\Delta G = T\,\Delta S - \Delta H$$

or

$$\Delta G = \Delta H - T\,\Delta S.$$

Thus, the change in the free energy of a system, for any constant-temperature process, is equal to the change in the enthalpy minus the product of the absolute temperature and the change in the entropy.

The property of free energy, changes of which are given by equation is defined by

$$G = H - TS$$

Since H and S are properties of the system, so is G.

FREE ENERGY FUNCTION

Free energy is the property that at any temperature, directly connects thermodynamics to chemical equilibria. The values of the free energies can be developed from calorimetric measurements or in some cases from the properties of the molecules of the substances.

Free Energies from Calorimetric Data

Tables of free-energy values at various temperatures can be built from the tabulated values at 25°C if heat-capacity data and, for any pertinent phase transitions, the enthalpy values are available. The usual procedure begins with the formal expression $G^o = H^o - TS^o$ for the free energy of a substance at some temperature. This relation can be developed so that it involves determinable quantities by subtracting the enthalpy of the substance at the reference temperature 298 K. This gives

$$G^o - H^o_{298} - H^o - H^o_{298} - TS^o.$$

Division by T leads to the less temperature-dependent quantities of the equation

$$\frac{G^o - H^o_{298}}{T} = \frac{H^o - H^o_{298}}{T} - S^o$$

The two terms on the right can be evaluated if calorimetric data are available over the temperature range from 298 K to the temperature of interest.

The quantity $(G^o - H^o_{298})/T$ is known as *free-energy function*. Values for this function are tabulated for various temperatures as illustrated in Appendix Table B-5. Since this free-energy function has negative values at all temperatures above 298 K, it is customary to present $-(G^o - H^o_{298})/T$ and to use joules rather than the usual kilojoules as the energy unit.

Free Energy and Molecular Properties

The standard free energy is given by the defining equation $G^o = H^o - TS^o$. Now we make use of our earlier molecular interpretations of the enthalpy and entropy of ideal-gas substances. We had

$$H^o - U^o = U^o - U^o_0 + RT \qquad \text{[ideal gas]}$$

and

$$S^o = \frac{U^o - U^o_0}{T} + R \ln \frac{q^o}{N} + R \qquad \text{[ideal gas]}$$

These equations, with $G^o = H^o - TS^o$, lead us to

$$Go - U^o_0 = - RT \ln\frac{q^o}{N}$$

and

$$G^o = U^o_0 - RT \ln\frac{q^o}{N} \qquad \text{[ideal gas]}$$

You now can see the molecular basis of the free-energy property. The standard free energy G^o is equal to the energy U^o_0 that the system would have if all the molecules were in their lowest-energy state modified by a term RT ln (q^o/N), which involves the partition function.

From this molecular interpretation, you can see why the free energy is a measure of the potential for "making things happen." If, for example, the partition function of a substance is large, the free energy of the substance is greatly reduced below the U^o_0 value. The molecules of such a substance can exist in many ways, and the substance has less tendency to react than we would have expected on the basis of the U^o_0 value.

Free-Energy Function from Molecular Properties

The thermal contribution to the standard-state free energy of an ideal gas is $G^o - U^o_0 = - RT \ln (q^o/N)$,. We saw how the partition function could be calculated for substances that can be treated as ideal gases. Such calculations can be carried out for conditions, high temperature for example, for which direct thermodynamic measurements are difficult.

Equilibrium Constants from Free-Energy Functions

The change in free energy that accompanies a chemical reaction can be calculated from the free-energy function values, such as those of Appendix Table B-5, which are based on 298-K enthalpy values or from functions such as those of Appendix Table B-4, which are based on 0-K enthalpy, or energy, values, or from both types of free-energy functions.

The basic relations that can be applied at any temperature are

$$\frac{\Delta G^o}{T} = \frac{\Delta(G^o - H^o_{298})}{T} + \frac{\Delta H^o_{298}}{T}$$

or

$$\frac{\Delta G^o}{T} = \frac{\Delta(G^o - U^o_0)}{T} + \frac{\Delta U^o_0}{T}$$

The ΔH^o_{298} values that are needed can often be calculated from values, such as those of Appendix Tables B-1 and B-2, of standard enthalpies of formation at 298 K. The U^o_0 values for the second formulation can also be calculated from standard enthalpies, or energies, of formation at 0 K. Such values are included in Appendix Tables B-3 and B-4. If values for the free-energy function must be drawn from the two types of free-energy functions, conversions can be made by using the relation,

$$G^o - H^o_{298} = (G^o - U^o_0) - (H^o_{298} - U^o_0)$$

or

$$G^o - U^o_0 = (G^o - H^o_{298}) + (H^o_{298} - U^o_0).$$

FREE ENERGY OF REAL GASES

For ideal gases, the standard-state molar free energy G^o is chosen as that of the gas at 1-bar pressure. The variation of free energy from the value for the standard state has been expressed for ideal gases by the formula

$$G = G^o + RT \ln P \qquad \text{[P in bar]}$$

We now need a procedure for choosing a standard state for real gases and for expressing the free energy of such gases as the conditions vary from this standard state.

For real gases we choose the standard state as that at which the fugacity would be equal to 1 bar if the gas followed ideal behaviour from zero pressure up to this fugacity. This selection of a hypothetical state for the standard state of non-ideal gases can be appreciated. Although, the standard state is not a real physical state, the free energy of the gas in its standard state can be deduced and used.

Suppose the free energy of the gas can be deduced for the gas at some pressure P up to which the gas behaves ideally. At that pressure we have f = P. Then the standard-state free energy G^o can be calculated directly from the relation

$$G^o - G = RT \ln \frac{f^o}{f}$$

or, with $f^o = 1$ and $f = P$

$$G^o = G - RT \ln P$$

Activity Coefficient

As we know,

$$G - G^o = RT \ln P$$

suggests that introduction of a term to exhibit the nonideality of a gas by comparing f with P. The activity *coefficient* γ, defined as

$$\gamma = \frac{f}{P}$$

is introduced. With this factor the free-energy expression for real gases can be written

$$G - G^o = RT \ln \gamma P$$

or $$G = G^o + RT \ln P.$$

The special convenience of y is that it shows explicitly the importance of non-ideality.

INTERRELATION OF THERMODYNAMIC PROPERTIES

All the properties of a chemical system, a sample of a substance or a mixture of substances have some fixed, definite values, when the state of the system is set by the selection of, for examples a temperature and pressure.

The properties that we have been dealing with have the symbols, V, U, H, S and G. These properties are all interrelated. As we know by thinking of the defining equations such as

$$H = U + PV$$

and $$G = H - TS.$$

Suppose the state of the system is changed the values of the properties of the system changes. These property changes must be interrelated.

Pressure and Volume Dependence of U

For any process, the change in the energy dU of the system is related to the change in the energies of the thermal and mechanical surroundings by

$$dU = -dU_{therm} - dU_{mech}$$

For a process in which only the mechanical energy is involved, $dU_{mech} = P\,dV$. For a reversible process $dU_{therm} = -T\,dS$. By considering this special process we arrive at the relation

$$dU = T\,dS - PdV.$$

Division of above equation by dP followed by specification of constant temperature gives

$$\left(\frac{\partial U}{\partial P}\right)_T = T\left(\frac{\partial S}{\partial P}\right)_T - P\left(\frac{\partial V}{\partial P}\right)_T$$

The pressure dependence of entropy given by equation can be inserted to give

$$\left(\frac{\partial U}{\partial P}\right)_T = T\left(\frac{\partial V}{\partial T}\right)_P - P\left(\frac{\partial V}{\partial P}\right)_T$$

The corresponding dependence of internal energy on volume can first writing the relation

$$\left(\frac{\partial U}{\partial V}\right)_T \left(\frac{\partial U}{\partial P}\right)_T\left(\frac{\partial P}{\partial V}\right)_T = -T\left(\frac{\partial V}{\partial T}\right)_P\left(\frac{\partial P}{\partial V}\right)_T - P\left(\frac{\partial V}{\partial P}\right)_T\left(\frac{\partial P}{\partial V}\right)_T$$

$$= -T\left(\frac{\partial V}{\partial T}\right)_P\left(\frac{\partial P}{\partial V}\right)_T - P$$

The $(\partial V/\partial T)_P$ term can be expressed from $dV = (\partial V/\partial T)_P\,dT + (\partial V/\partial P)_P\,dP$ by specifying constant volume, and rearranging to

$$\left(\frac{\partial V}{\partial T}\right)_P = -\left(\frac{\partial V}{\partial P}\right)_P\left(\frac{\partial P}{\partial T}\right)_V$$

Now the equation for $(\partial U/\partial V)_T$ becomes

$$\left(\frac{\partial U}{\partial V}\right)_T = T\left(\frac{\partial P}{\partial T}\right)_V = -P.$$

Energy of an Ideal Gas

The internal energy U of a sample of an ideal gas depends on only the temperature, not on the pressure or volume of the sample.

This idea was justified by the kinetic-molecular theory We can also show that it holds withouts stepping out of the classical thermodynamics.

We can use conformity to the equation PV = nRT as a definition of ideal gas behaviour,

we can write,

$$\left(\frac{\partial U}{\partial P}\right)_T = 0$$

and

$$\left(\frac{\partial U}{\partial V}\right)_T = 0$$

Thus, without any stipulation other than PV = nRT, we arrive at the conclusion that the internal energy of an ideal gas depends on only the temperature.

Relation between C_P and C_V

Now we can return to the comparison of C_P and C_v

$$C_P - C_V = \left[P + \left(\frac{\partial U}{\partial V}\right)_T\right]\left(\frac{\partial V}{\partial T}\right)_P$$

Now we can use equation to replace $(\partial U/\partial V)_T$. This replacement leads to

$$C_P - C_V = \left(\frac{\partial P}{\partial T}\right)_V \left(\frac{\partial V}{\partial T}\right)_P$$

(Notice that the ideal-gas equation of state PV = RT would again lead to the result $C_P - C_V = R$.)

The pressure-temperature coefficient $(\partial V/\partial T)_V$, is conveniently replaced by volume-pressure and volume-temperature coefficients. The total differential for the volume can be expressed as

$$dV = \left(\frac{\partial P}{\partial T}\right)_P dT + \left(\frac{\partial V}{\partial P}\right)_T dP$$

Stipulation of constant volume gives, after rearrangement,

$$\left(\frac{\partial P}{\partial T}\right)_V = \frac{\left(\frac{\partial V}{\partial T}\right)_P}{\left(\frac{\partial V}{\partial P}\right)_T}$$

Then equation can be recast as

$$C_P - C_V = -\frac{T\left(\frac{\partial V}{\partial T}\right)_P^2}{\left(\frac{\partial V}{\partial P}\right)_T}$$

This can be expanded

$$\frac{PV}{RT} = 1 + \left(b - \frac{a}{RT}\right)\frac{1}{V} + \frac{b^2}{V^2} + \ldots$$

If only the first of the variable terms on the right is retained and V is replaced by RT/P in this term,

we get

$$V = \frac{RT}{P} + b - \frac{a}{RT}$$

The derivations of previous equations, can be shown as follows:

$$C_P - C_V = R\left(1 + \frac{2a}{R^2T^2}P\right)$$

For liquids and solids, thermal expansively α defined by,

$$\alpha = \frac{1}{V}\left(\frac{\partial V}{\partial T}\right)_P$$

and isothermal compressibility β, defined by

$$\beta = -\frac{1}{V}\left(\frac{\partial V}{\partial T}\right)_P$$

with these expressions we can write

$$C_P - C_V = \frac{\alpha^2 VT}{\beta}.$$

FREE ENERGY AND TEMPERATURE

The way in which free energy depends on temperature shows the way in which the equilibrium constant depends on temperature.

The equilibrium constant can be evaluated from the relation $AG^\circ = -RT \ln K$ at any temperature for which free-energy data are available. Frequently we calculate the equilibrium constant at 25°C, the reference temperature for which most thermodynamic data are tabulated. Now an expression for the temperature dependence of the equilibrium constant is developed.

It will allow us to extend an equilibrium-constant value to other temperatures, and it will show on what the temperature variation of an equilibrium constant depends.

The free energy of each substance involved in a reaction depends on the temperature,

$$\left(\frac{\partial G}{\partial T}\right)_P = -S$$

We want to use this relation to deal with the temperature dependence of the equilibrium constant. This constant is related to free energies through the relation $AG^\circ = -RT\ln K$.

Recall that the symbol AG and a corresponding ΔS symbol designate

$$\Delta G = G_{prod} - G_{react}$$

and
$$\Delta S = S_{prod} - S_{react}.$$

We have the relation

$$\left[\frac{\partial(\Delta G)}{\partial T}\right]_P = -\Delta S$$

At any constant temperature the changes of free energy, enthalpy, and entropy for any reaction are related by

$$\Delta G = \Delta H - T\,\Delta S$$

or
$$\Delta S = \frac{\Delta H - \Delta G}{T} \qquad [T \text{ const}]$$

The second expression can be used to eliminate ΔS to give

$$\left[\frac{\partial(\Delta G)}{\partial T}\right]_P = \frac{\Delta H - \Delta G}{T} = -\frac{\Delta H}{T} + \frac{\Delta G}{T}$$

or $$\left[\frac{\partial(\Delta G)}{\partial T}\right]_P - \frac{\Delta G}{T} = \frac{\Delta H}{T}$$

The two terms on the left side of equation can be shown to be equivalent to

$$T\left[\frac{\left(\frac{\partial \Delta G}{T}\right)}{\partial T}\right]_P = T\frac{T\left[\partial\left(\frac{\Delta G}{\partial T}\right)\right]_P - \Delta G}{T^2} = \left[\partial\frac{(\Delta G)}{\partial T}\right]_P = \frac{\Delta G}{T}$$

Now the left side of equation can be inserted, then we get

$$T\left[\frac{\left(\frac{\partial \Delta G}{T}\right)}{\partial T}\right]_P = \frac{\Delta H}{T}$$

Finally, standard states are indicated and the relation $\Delta G^o = -RT \ln K$ inserted to give, on rearrangement,

$$\frac{d(\ln K)}{dT} = \frac{\Delta H^o}{RT^2}$$

This important formula is the goal of the derivation. The rate of change of the equilibrium constant with temperature is seen to depend on the standard enthalpy change. The change of ln K or K can be obtained by an integration of this expression either with the assumption of a constant value of ΔH^o or with the temperature dependence of this quantity expressed by the empirical expressions

$$\frac{d(\ln K)}{d\left(\frac{1}{T}\right)} = -\frac{\Delta H^o}{R}$$

The integrated form of these equations, on the assumption that ΔH temperature-independent, is

$$\ln K = -\frac{\Delta H^o}{RT} + \text{const}$$

CONCEPT OF HYDROLYSIS

The reaction of an anion or cation with water accompanied by cleavage of O—H bond is called Hydrolysis.

The term hydrolysis is derived from *hydro*, meaning water, and *lysis*, meaning breaking. It may be noted that in *anionic hydrolysis* the solution becomes slightly basic (pH > 7) due to the generation of excess OH ions. In *cationic hydrolysis*, there is excess of H^+ ions which makes the solution slightly acidic (pH < 7).

Bronsted-Lowry Concept of Hydrolysis

$$\underset{\text{weak acid}}{HA} + H_2O \rightleftharpoons H_3O^+ + \underset{\text{conjugate base}}{A^-}$$

Since HA is a weak acid (poor proton donor), its conjugate base, A^-, must be relatively strong (good proton acceptor). Owing to this fact. A^- ions tend to react with water by accepting proton from the latter to form HA molecules (anionic hydrolysis),

$$A^- + H_2O \rightarrow HA + OH^-$$

The presence of OH' ions makes the solution basic.

Similarly, BOH and B^+ are a conjugate acid-base pair. Since BOH is a weak base, its conjugate acid, B^+, would be relatively strong. Thus B^+ would accept OH^- ions from water to form BOH molecules.

$$B^+ + H_2O \rightarrow BOH + H^+$$

The presence of excess H^+ ions makes the solution acidic.

NaCl dissociates in water to give the anion Cl^- and HCl and Cl^- constitute an acid-base conjugate pair.

$$HCl + H_2O \rightleftharpoons H_3O + Cl^-$$

conjugate base

Since HCl is a strong acid, Cl^- is very weak base. Cl^- is unable to accept a proton (H^+) from an acid, particularly water. That is why Cl^- does not hydrolyse. It can not generate OH^- ions as follows:

$$Cl + H_2O \not\rightarrow HCl + OH^-$$

The pH of sodium chloride solution remains unaffected.

The different salts may be classified into the following types according to their hydrolytic behaviour.

(1) Salts of Weak acids and Strong bases

(2) Salts of Weak bases and Strong acids

(3) Salts of Weak acids and Weak bases.

We have already considered that a salt of a strong acid and strong base *e.g.*, NaCl, does not show hydrolysis.

Sodium acetate, CH_3COONa, and sodium cyanide, NaCN, are examples of this type of salts.

Sodium Acetate CH_3COONa

This is a salt of the weak acid, CH_3COOH, and strong base,

NaOH. It ionises in aqueous solution to form the anion CH_3COO^-. Being the conjugate base of a weak acid, CH_3COOH, it is a relatively strong base. Thus CH_3COO^- accepts H^+ ion from water and undergoes hydrolysis.

$$CH3COO^- + H_2O \rightarrow CH3COOH + OH^-$$

The resulting solution is slightly basic due to excess OH" ions present

Sodium Cyanide, NaCN

It is the salt of a weak acid, HCN, and NaOH. It ionises to form CN^- anions. Being conjugate base of a weak acid, CN' is relatively strong base. Thus the anion CN^- accepts a H^+ ion from water and undergoes hydrolysis.

$$CN^- + H_2O \rightleftharpoons HCN + OH^-.$$

The solution becomes basic due to the generation of OH^- ions.

Some salts of weak bases and strong acids undergo cationic hydrolysis and yield slightly acidic solutions.

Ammonium chloride is a typical example of this class of salts. It is the salt of a weak base.

NH_4OH, and strong acid, HCl. It ionises in aqueous solution to form the cation, NH_4^+.

$$NH_4OH \rightleftharpoons \underset{\text{conjugate acid}}{NH_4^+} + OH^-$$

NH_4^+ is a Bronsted conjugate acid of the weak base NH_4OH. Therefore, it is a relatively strong acid. It accepts OH^- ion from water (H_2O) and forms the unionised NH_4OH and H^+ ion.

$$NH_4^+ + H_2O \rightleftharpoons NH_4OH + H^+$$

The accumulation of H^+ ions in solution makes it acidic.

Relation Between Hydrolysis Constant and Degree of Hydrolysis

The degree of hydrolysis is the fraction of the salt which has undergone hydrolysis when equilibrium is established. It is generally represented by α.

Suppose we start with one mole of the salt dissolved in V litres of solution. Then the equilibrium concentrations are:

$$A^- + H_2O \rightleftharpoons HA + OH^-$$

Equilibrium concentrations $\quad \frac{1-\alpha}{V} \qquad \frac{\alpha}{V} \qquad \frac{\alpha}{V}$

Hence the hydrolysis constant A), is given by

$$K_h = \frac{[HA][OH^-]}{[A^-]} = \frac{\frac{\alpha}{V} \times \frac{\alpha}{V}}{\frac{1-\alpha}{V}} = \frac{\alpha^2}{(1-\alpha)V}$$

If α is small, $(1 - \alpha)$ may be taken as equal to one. Then,

$$K_h = \frac{\alpha^2}{V}$$

$$\alpha^2 = K_h V = \frac{K_w}{K_a} V$$

$$\alpha = \sqrt{\frac{K_w V}{K_a}}$$

$$= \sqrt{\frac{K_w}{K_a C}}.$$

where C is the initial concentration of the salt. Knowing the values of K_w, K_a and C, the degree of hydrolysis can be calculated.

HYDROLYSIS CONSTANT

Applying the law of mass action to a hydrolysis reaction, the hydrolysis constant K_h is given by

$$K_h = \frac{[H^+][BOH]}{[B^+][H_2O]}$$

Since $[H_2O]$ is very large it is taken to be constant and the hydrolysis constant expression is reduced to,

$$K_h = \frac{[H^+][BOH]}{[B^+]}$$

Relation Between K_h, K_w and K_b

We know that the ionic product of water K_w is expressed as :

$$K_w = [H^+][OH^-]$$

For the dissociation of a weak base, BOH,

$$BOH \rightleftharpoons B^+ + OH^-$$

the dissociation constant, K_b can be expressed as :

$$K_b = \frac{[B^+][OH^-]}{[BOH]}$$

$$\frac{K_w}{K_b} = \frac{[H^+][BOH]}{[B^+]} = K_h$$

or $$\frac{K_w}{K_b} = K_h.$$

Thus, the hydrolysis constant, K_h, varies inversely as the dissociation constant, K_b of the base. Therefore *weaker the base greater will be the hydrolysis constant of the salt.*

Relation Between Hydrolysis Constant and Degree of Hydrolysis

Suppose we start with one mole of the salt dissolved in V litres of solution. Then the concentrations when equilibrium is attained are:

$$B^+ + H_2O \rightleftharpoons BOH + H^+$$

Equilibrium concentrations $\quad \frac{1-\alpha}{V} \qquad \frac{\alpha}{V} \qquad \frac{\alpha}{V}$

Applying the Law of Mass Action, the hydrolysis constant, K_h is given by the expression

$$K_h = \frac{[H^+][BOH]}{[B^+]} = \frac{\frac{\alpha}{V} \times \frac{\alpha}{V}}{\frac{1-\alpha}{V}} = \frac{\alpha^2}{(1-\alpha)V}$$

When α is small, $(1 - \alpha)$ may be considered as equal to one. Then we have

$$K_h = \frac{\alpha^2}{V}$$

or $$K_h \times V = \alpha^2$$

or $$\alpha = \sqrt{K_h \times V}$$

$$K_h = \frac{K_w}{K_b}$$

$$\alpha = \sqrt{\frac{K_w}{K_b} \times V}$$

$$= \sqrt{\frac{K_w}{K_b \times C}}$$

where C is the initial concentration of the salt.

Derivation of pH

From the above discussion it is clear that

$$[H^+] = \frac{\alpha}{V} = \alpha \times C$$

Substituting the value of a from equation (5), we have

$$[H+] = \frac{1}{V}\sqrt{\frac{K_w \times V}{K_b}} = \sqrt{\frac{K_w}{K_b V}} = \sqrt{\frac{K_w \times C}{K_b}}$$

Taking logarithms and reversing 'he signs

$$-\log[H^+] = -\frac{1}{2}\log K_w - \frac{1}{2}\log C + \frac{1}{2}p\,K_b$$

or $$pH = 7 + \frac{1}{2}p\,K_b - \frac{1}{2}\log C$$

In this case it is evident that pH will always the less than 7. Thus, *the solution of a salt of weak base and strong acid will always be acidic.*

DETERMINATION OF DEGREE OF HYDROLYSIS

The degree hydrolysis of a salt can be determined by a number of methods. The more important ones are described below.

Dissociation Constant Method

The degree of hydrolysis, α, is related to the ionic product of water, K_w and the dissociation constant of the weak acid, K_a, or of the weak base, K_b, from which the salt is obtained The various relationships are listed below.

For salt of a Weak acid and Strong base:

$$\alpha = \sqrt{\frac{K_w}{K_a \times C}}$$

For salt of a Weak base and Strong acid:

$$\alpha = \sqrt{\frac{K_w}{K_b \times C}}$$

For salt of a Weak acid and Weak base:

$$\alpha = \sqrt{\frac{K_w}{K_a \times K_b}}$$

Substituting the values of K_w, K_a, K_b, and C, the initial concentration of the salt, in the appropriate expression, a can be calculated.

It may be noted that the degree of hydrolysis for the salt of a weak acid and weak base is independent of the concentration. However in this case, the value of α is not small and $(1 - \alpha)$ cannot be taken as equal to one. Therefore the relationship for calculating the degree of hydrolysis is considered in the form

$$\frac{\alpha^2}{(1-\alpha)^2} = \sqrt{\frac{K_w}{K_a \times K_b}}$$

This is by far the most accurate method for determining the degree hydrolysis of a salt and is used n all modem work.

From Conductance Measurements

The degree of hydrolysis, α, of a salt can be determined by conductance measurements. Let us consider a solution containing the salt of a weak base and a strong acid. The hydrolysis reaction can be written as

$$\underset{(1-\alpha)}{B^+} + H_2O \rightleftharpoons \underset{\alpha}{BOH} + \underset{\alpha}{H^+}$$

If it be assumed that the base is so weak that it is not dissociated at all, it will contribute nothing to conductance of the solution. The equivalent conductance of the salt, therefore, consists of :

that due to $(1 - \alpha)$ equivalents of the salt.

that due to a equivalents of the acid produced by hydrolysis.

Thus, we can say that:

$$\Lambda = (1 - \alpha)\, \Lambda_{salt} + \alpha \Lambda_{acid}$$

$$\Lambda - \Lambda_{salt} = \alpha\, (\Lambda_{acid} - \Lambda_{salt})$$

or

$$\alpha = \frac{\Lambda - \Lambda_{salt}}{(\Lambda_{acid} - \Lambda_{salt})}$$

Λ is found by conductance measurements. Λ_{acid} is taken as the value for strong acid at infinite dilution. Λ_{acid} is determined by adding excess of weak base to the solution to suppress hydrolysis so that the resulting experimentally determined value of Λ can be taken as that of the unhydrolysed salt.

ELECTROCHEMICAL CELLS

An electrical connection to a solution can be made with several types of electrodes.

The reactions that occur in electrochemical cells generally involve the ionic species of parent electrolytes. Some of these are responsible for the passage of the electric current through the usually aqueous solution in the cell. Information on the free energies of all the species, ionic or not, involved in the cell reaction is obtained from the results of electrical measurements.

Gas Electrodes

A gas can be induced to participate in an electrochemical reaction by means of an electrode like. An example that is important in the development of electrochemical data is the hydrogen electrode. The electrode is such that on the surface of the inert-metal electrode the reagents $H_2(g)$, $H^+(aq)$, or $H_3O^+(aq)$ and e^-, the latter in the metallic conductor, can be accommodated. The electrode reaction that can then proceed is

$$H^+ - e^- \rightleftharpoons H_2$$

The electrode reaction is written, as is the convention, as an electron-acquiring reduction reaction. The electrode assembly in which this reaction.

$$Pt \mid H_2(P\ bar) \mid H^+c\ mol\ L^{-1})$$

The vertical lines indicate interfaces between physical states, and the significant features of these states are indicated. The symbol Pt is more restricted than necessary, being used here to imply any inert-metal electrode.

Amalgam Electrodes

In a variation of the previous electrode, the metal is in the form of an amalgam; *i.e.*, it is dissolved in mercury, rather than in the pure form Electrical contact is made by a platinum wire dipping into the amalgam pool. The reaction is the same as in the metal-metal-ion electrode, with the mercury playing no chemical role.

The particular value of amalgam electrodes is that active metals such as sodium can be used in such electrodes. A sodium-amalgam electrode is represented as

$$Na\ (\text{in Hg at } c_1)\ |\ Na^+(c_2)$$

where the concentration of the sodium metal in the mercury as well as that of the sodium ion in the solution must be given. In addition to allowing the study of active metals, the amalgam electrode illustrates some of the thermodynamic relations which will be obtained for electrochemical cells.

Metal-Insoluble-Salt Electrodes

A more elaborate but usually satisfactory and frequently used electrode consists of a metal in contact with an insoluble salt of the metal, which in turn is in contact with a solution containing the anion of the salt. An example is represented as

$$Ag\ |\ AgCl\ |\ Cl^-(c)$$

The electrode process at such an electrode is

$$AgCl(s) \rightleftharpoons Ag^+ + Cl^-$$

$$Ag^+ + e^- \rightleftharpoons Ag$$

$$AgCl(s) + e^- \rightleftharpoons Ag + Cl$$

The electrode reaction involves only the connection of Cl^- as a variable in contrast with the Ag/Ag^+ electrode which has the concentration as a variable.

EMF and Free Energy

The emf of an electrochemical cell can be measured and can be attributed to the chemical reaction that occurs.

When pairs of electrodes, are combined and are connected by an external electrical conductor, an electric current flows and chemical reactions occur in the two half-cells. Electrodes can be combined without complication if both electrodes can operate in the same solution. When this is not tolerated, a connection must be made between the solutions that allows ionic conduction to occur between the half-cells but prevents mixing of the two half-cell solutions. A KCl *salt bridge*, illustrated in the assembly of often used. Such a coupling device, indicated by a double vertical line in a cell diagram, has no net influence on the cell reaction. Although some difficulties are introduced into the quantitative interpretation of the cell voltage, or *electromotive force* (emf).

The voltage, or emf, of a cell can be measured by placing a voltmeter across the terminals; However, the measured voltage would depend on the current drawn off by the voltmeter. The maximum voltage is that produced in the limit as zero current is drawn. In practice, a suitable value for this characteristic of the cell can be obtained with a vacuum-tube voltmeter (VTVM), which draws a very small current or, more accurately, by a potentiometer arrangement.

The free energy change for the cell reaction is related to the cell e.m.f.

When the e.m.f. of a cell is measured with the reaction proceeding at a state of balance, or reversibly the mechanical energy that can be obtained from the cell current is the maximum energy, over and above any P dV energy, that can be delivered to the mechanical surroundings.

The electrical energy produced by a cell as it drives 1 mol of electrons through an opposing voltage is equal to the product of the total charge of these electrons and the voltage through which this charge is driven. The charge of 1 mol of electrons is 96,485 C, or 1 **F**. With this notation, the mechanical energy producted as n mol of electrons is driven by the cell emf E through a nearly balancing voltage is

$$\Delta U_{mech} = n\mathbf{F}\,\mathbf{E}$$

This energy, ΔU_{mech}, is the mechanical energy, other than P dV type of energy, that the reaction produces when it proceeds reversibly. This energy is equal to the *decrease* in the free energy of the reacting system as a result of the process described by the cell reaction. Thus we have

$$\Delta G = n\,\mathbf{F}\,\mathbf{E}$$

where **E** is the reversible emf of the cell. Notice that the sign of ΔU_{mech} thus of ΔG, is such that for the direction of spontaneous change, which is associated with a positive emf, we do obtain the necessary negative ΔG. As an example, for the cell reaction of equation and the E° value of +0.2223 V, we have

$$AgCl + H_2 \rightarrow Ag + H^+ + Cl^-$$

$$\Delta G^\circ = -(1)(96{,}485)(+0.2223) = -21{,}450 \text{ J} = -21.45 \text{ kJ}$$

If we assume a basic electrolyte, we could have described the electrode reactions as

Right reaction: $O_2 + H_2O(l) + 2e^- \rightarrow 2OH$

Left reaction: $H_2(g) + 2OH^- \rightarrow 2H_2\,O(l) + 2e^-$

Cell reaction: $H_2(g) + O_2(g) \rightarrow H_2O(l)$.

Notice that the acid-base character of the electrolyte does not affect the electron changes or the overall reaction. In both cases n = 2. The mechanical energy that can be produced is

$$\Delta U_{mech} = n\mathbf{F}\,\mathbf{E}$$

$$= 2(96{,}485C)(1.229 \text{ V}) = 237{,}200 \text{ J} = 237.2 \text{ kJ}$$

EMF AND ACTIVITIES

The dependence of the emf of a cell on the concentration, or more directly on the activity, of the variable reagents can be calculated from our knowledge of the relation between the free energy and the activity. Consider an electrochemical cell for which the overall chemical reaction is

$$aA + bB \rightleftharpoons cC + dD.$$

Further, suppose that A, B, C, and D are reagents whose concentration can be varied; *i.e.*, they are gases or solutes in aqueous solution. If, in addition, there are solids involved in the reaction, they will contribute only a constant term to the results, as we shall see.

For species A the free energy of 1 mol can be written from the treatment of solutes.

$$G_A = G_A^\circ + RT \ln a_A.$$

The free-energy per mole is written with the overbar which indicates, a partial molal free energy. If the reagent is a pure substance rather than the component of a solution, the partial molal free energy is just the free energy of 1 mol of the substance.

For a mol of reagent A, the free energy is

$$aG_A = aG_A^\circ + aRT \ln a_A = aG_A^\circ + RT \ln (a_A)^a$$

From expressions like these for all four reagents, the free-energy change for the overall cell reaction can be deduced,

$$\Delta\overline{G} = \Delta\overline{G}° + RT \ln \frac{(a_C)^c (a_D)^d}{(a_A)^a (a_B)^b}$$

where $\Delta G°$ is the difference in free energy of the products and the reactants when all reagents are in their standard states. Furthermore,

$$A\overline{G} = -nF\,\mathbf{E} \quad \text{and} \quad A\overline{G}° = -nF\,\mathbf{E}°$$

which gives

$$\mathbf{E} = \mathbf{E} - \frac{RT}{nF} \ln \frac{(a_C)^c (a_D)^d}{(a_A)^a (a_B)^b}$$

This expression for the activity dependence, or approximately the concentration dependence, of emf is known as the *Nernst equation.*

This important equation shows how the emf of a cell can be calculated from the emf for all reagents in their standard states and the activities of the reagents.

It is clear that when the activities of all the reagents are unity, the logarithmic term drops out and $\mathbf{E} = \mathbf{E}°$, that is, the emf for the standard states.

At 25°C, the temperature at which most standard electrode potentials are reported, the factor before the logarithm term can be explicitly worked out and gives

$$E = E° - \frac{0.05915}{n} \log \frac{(a_C)^c (a_D)^d}{(a_A)^a (a_B)^b}$$

Included in the numerical factor is the term 2.3026 for the conversion from natural logarithms to logarithms to the base 10.

Note that the activity term has the familiar form of the equilibrium-constant expression. The activities of the reagents, however, have the values that are determined by the solutions used to make up the cell. These are not generally the equilibrium values. As in the development of the equilibrium-constant expression, the activities of solid or otherwise fixed-concentration reagents are not explicitly included.

ELECTROLYTE-CONCENTRATION

Another special type of cell is that in which the cell emf depends only on the electrolyte concentration.

A second type of cell whose emf is derived only from the free-energy change of a dilution reaction is that in which the electrolyte of the cell is involved in the dilution. If one attempts to cônstruct such a cell by having two solutions of different concentrations in physical contact with each other, complications arise because of the non-equilibrium processes that occur at the liquid-liquid junction.

For the present only the simpler *concentration cells without liquid junctions* are considered. Such cells can be illustrated with two cells of the type

$$Pt \mid H_2 \mid HCl(c) \mid AgCl(s) \mid Ag$$

each of which has a reaction

$$H_2 + AgCl \rightleftharpoons Ag + H^+(c) + Cl^-(c)$$

Consider two such cells electrically connected through their silver electrodes in the opposed manner

$$Pt \mid H_2 \mid HCl(c_1) \mid AgCl \mid Ag\text{–}Ag \mid AgCl \mid HCl(c_2) \mid H_2 \mid Pt.$$

The overall reaction is now the sum of the two simple cell reactions. If the pressure of hydrogen gas is the same for both terminal electrodes, the electrode reactions can be combined to give

Right cell: $Ag + H^+(c_2) + Cl^-(c_2) + e^- \rightleftharpoons H_2 + AgCl$

Left cell: $H_2 + AgCl \rightleftharpoons Ag + H^+(c_1) + Cl^-(c_1) + e^-$

Overall reaction: $H^+(c_2) + Cl^-(c_2) \rightleftharpoons H^+(c_1) + Cl^-(c_1)$

The overall reaction therefore involves no chemical change and consists only of the transfer of HCl from a concentration c_2 to a concentration c_1. The emf of the complete cell is expressed as

$$E = \frac{0.05915}{1} \log \frac{(a_{H^+})_1 (a_{Cl^-})_1}{(a_{H^+})_2 (Cl^-)_2} = -0.05915 \log \frac{[a_\pm]_1^2}{[a_\pm]_2^2}$$

$$= 0.011830 \log \frac{(a_\pm)_1}{(a_\pm)_2}$$

Again one can see that the spontaneous process takes HCl from a higher activity, or concentration, to a lower activity, or concentration. If, for example, c_2, is greater than c_1 and therefore $(a_\pm)_2$ is greater than $(a_\pm)_1$, the emf will be positive and the reaction will proceed in the direction in which it is written.

Such concentration cells can be used to determine the activity of an electrolyte at one concentration or in a solution containing other ions compared with the activity of an electrolyte in another solution.

The principal purpose in presenting these concentration cells here, however, is to provide a contrast tot eh situation that arises when concentration cells with liquid junctions are dealt with.

ION-SELECTIVE MEMBRANE

An ion-selective membrane can be used to form an electrochemical cell whose emf depends on the concentration of that ion.

Now let us study in briefly about, the glass electrode and calomel electrode of a pH meter.

The glass membrane of the glass electrode separates two different solutions. It provides for general conduction across the liquid junction. The glass membrane most often used leads to a cell whose emf is primarily responsive to hydrogen ions. Glasses can be made that allow passage of only one type of ion, in this case the hydrogen ion, and thus the electrode can be constructed to be sensitive to this ion only. Much of the importance of the glass electrode stems from its lack of response to various oxidizing and reducing agents and to a large variety of ionic species. Difficulties may occur, however, if the glass electrode is used in solutions of high sodium-ion concentration or in solution sufficiently alkaline to attack the glass membrane.

The glass electrode that responds to variations in the hydrogen activity is just one member of a growing list of practical electrode devices known as *ion-selective-membrane electrodes*. The three general types Modifications of the glass membrane of a glass electrode can make the membrane permeable and the electrode responsive to a variety of ions. Glass electrodes have been prepared that are sensitive to each of the ions of the alkali-metal family and to other ions such as NH_4^+, Ag^+ and Cu^+ as well as H^+.

Solid-state membranes can be represented by the fluoride-ion electrode. The membrane material is the sparingly soluble crystalline substance LaF_3. Fluoride ions are conducted through the crystal with ease while other ions are rejected. The fluoride-ion electrode is highly specific and rugged, and thus is a valuable analytic tool.

Liquid-membrane electrodes can be varied by the choice of liquid, which can be organic or inorganic, and by the presence of dissolved ion-exchange substances. Liquid membranes have been developed for electrodes selective for Ca^{2+}, Mg^{2+}, NO_2^-, Cl^-, and many others.

When supplemented with a reference electrode, all such electrodes produce a potential that more or less conforms to the equation

$$E = \text{const} + \frac{RT}{nF} \ln a_M$$

where a_M is the activity of the ion to which the electrode is selective. In practice, measurements must be made of known solutions containing potentially interfering ions, to judge the conformity to this Nernst-type expression.

OSTWALD'S LAW

As we know an electrotype dissociates into ions in water solutions. These ions are in a state of equilibrium with the undissociated molecules. This equilibrium is called the Ionic equilibrium. Ostwald noted that the Law of Mass Action can be applied to the ionic equilibrium as in the case of chemical equilibria.

Let us consider a binary electrolyte AB which dissociates in solution to form the ions A^+ and B^-.

$$AB \rightleftharpoons A^+ + B^-$$

Let C moles per litre be the concentration of the electrolyte and a (alpha) its degree of dissociation. The concentration terms at equilibrium may be written as:

$$[AB] = C(1-\alpha) \text{ mol litre}^{-1}$$

$$[A^+] = C\alpha \text{ mol litre}^{-1}$$

$$[B^-] = C\alpha \text{ mol litre}^{-1}$$

Applying the Law of Mass Action

Rate of dissociation $= k_1 \times C(1-\alpha)$

Rate of combination $= k_2 \times C\alpha \times C\alpha$

At equilibrium:

$$k_1 \times C(1-\alpha) = k_2 \times C\alpha \times C\alpha$$

or $$\frac{C\alpha \times C\alpha}{C(1-\alpha)} = \frac{k_1}{k_2} = K_c$$

or $$K_c = \frac{\alpha^2 C}{(1-\alpha)} \text{ mol litre}^{-1}$$

The equilibrium constant K_C is called the *Dissociation constant* or *Ionization constant.* It has a constant value at a constant temperature.

If one mole of an electrolyte be dissolved in V litre of the solution, then

$$C = \frac{1}{V}$$

V is, known as the *Dilution* for the solution. Thus the expression (1) becomes

$$K_c = \frac{\alpha^2}{(1-\alpha)V}$$

This expression which correlates the variation of the degree of dissociation of an electrolyte with dilution, is known as Ostwald's Dilution Law.

For Weak Electrolytes

For weak electrolytes, me value of α is very small as compared to 1, so that in most of the calculations we can take $1 - \alpha \simeq 1$. Thus, the Ostwald's Dilution Law expression become?

$$K_c = \frac{\alpha^2}{V}$$

It implies that *the degree of dissociation of a weak electrolyte is proportional to the square root of the dilution i.e.,*

$$\alpha \propto \sqrt{K_c V}$$

or $$\alpha = K'\sqrt{V}.$$

Factors that explain the failure of Ostwald's law in case of strong electrolytes :

1. The law is based on Arrhenius theory which assumes that only a fraction of the electrolyte is dissociated at ordinary dilutions and complete dissociation occurs only at infinite dilution. However, this is true for weak electrolytes. Strong electrolytes are almost completely ionised at all dilutions and λ_v/λ_∞ does not give the accurate value of α.
2. The Ostwald' s law is derived on the assumption that the Law of Mass Action holds for the ionic equilibria as well. But when the concentration of ions is very high, the presence of charges

appreciately affects the equilibrium. Thus, the Law of Mass Action in its simple form cannot be applied.

3. The ions obtained by dissociation may get hydrated and may affect the concentration terms. Better results are obtained by using *activities* instead of concentrations.

DEGREE OF DISSOCIATION

When a certain amount of electrolyte (A^+B^-) is dissolved in water, a small fraction of it dissociates to form ions (A^+ and B^-). When the equilibrium has been reached between the undissociated and the free ions, we have

$$AB \rightleftharpoons A^+ + B^-.$$

The fraction of the amount of the electrolyte in solution present as free ions is called the Degree of dissociation.

If the degree of dissociation is represented by x, we can write

$$x = \frac{\text{amount dissociated (mol/L)}}{\text{initial concentration (mol/L)}}$$

The value of x can be calculated by applying the Law of Mass Action to the ionic equilibrium stated above:

$$K = \frac{[A^+][B^-]}{[AB]}.$$

If the value of the equilibrium constant, K, is given, the value of can be calculated.

Factors which Influence the Degree of Dissociation

The *degree of dissociation* of an electrolyte in solution depends upon the following factors :

Nature of Solute

The nature of solute is the chief factor which determines its degree of dissociation in solution. Strong acids and strong bases, and the salts obtained by their interaction are almost completely dissociated in solution. On the other hand, weak acids and weak bases and their salts are feebly dissociated.

Nature of the Solvent

The nature of the solvent affects dissociation to a marked degree. It weakens the electrostatic forces of attraction between the two ions and separates them. This effect of the solvent is measured by its *'dielectric constant'*. The dielectric constant of a solvent may be defined as *its capacity to weaken the force of attraction between the electrical charges immersed in that solvent.*

The dielectric constant of any solvent is evaluated considering that of vacuum as unity. It is 4.1 in case of ether, 25 in case of ethyl alcohol and 80 in case of water. The higher the value of the dielectric constant the greater is the dissociation of the electrolyte dissolved in it because the electrostatic forces vary inversely as the dielectric constant of the medium. Water, which has a high value of dielectric constant is, therefore, a strong dissociating solvent The electrostatic forces of attraction between the ions are considerably weakened when electrolytes are dissolved in it and as a result, the ions begin to move freely and there is an increase in the conductance of the solution.

Concentration

The extent of dissociation of an electrolyte is inversely proportional to the concentration of its solution. The less concentrated the solution, the greater will be the dissociation of the electrolyte. This is obviously due to the fact that in a dilute solution the ratio of solvent molecules to the solute molecules is large and the greater number of solvent molecules will separate more molecules of the solute into ions.

Temperature

The dissociation of an electrolyte in solution also depends on temperature. The higher the temperature greater is the dissociation. At high temperature the increased molecular velocities overcome the forces of attraction between the ions and consequently the dissociation is great.

SOLUBILITY EQUILIBRIA AND THE SOLUBILITY PRODUCT

When an ionic solid substance dissolves in water, it dissociates to give separate cations and anions. As the concentration of the ions in solution increases, they collide and reform the solid phase. Ultimately, a dynamic equilibrium is established between the solid phase and the cations and anions in solution. For example for a sparingly soluble salt, AgCl, we can write the equilibrium equation as follows:

$$\underset{\text{solid}}{AgCl} \rightleftharpoons Ag^+ + Cl^-$$

A Saturated solution is a solution in which the dissolved and undissolved solute are in equilibrium.

A saturated solution represents the limit of a solute's ability to dissolve in a given solvent This is a measure of the "solubility" of the solute.

The Solubility (S) of a substance in a solvent is the concentration in the saturated solution.

Solubility of a solute may be represented in grams per 100 mL of solution. It can also be expressed in moles per litre.

Molar Solubility is defined as the number of moles of the substance per one litre (L) of the solution.

The value of solubility of a substance depends on the solvent and the temperature.

Applying the Law of Mass Action to the above equilibrium for AgCl, we have

$$K = \frac{[A^+][B^-]}{[AgCl]}$$

The amount of AgCl in contact with saturated solution does not change with time and the factor [AgCl] remains the same. Thus the equilibrium expression becomes

$$K_{sp} = [Ag^+]\,[Cl^-]$$

where $[Ag^+$ and $[Cl^-]$ are expressed in mol/L. The equilibrium constant in the new context is called the *Solubility Product Constant* (or simply the *Solubility Product*) and is denoted by K_{sp}. The value of K_{sp}, for a particular solubility equilibrium is constant at a given temperature. The product $[Ag^+]\,[Cl^-]$ in the K_{sp} expression above is also called the *Ionic Product* or *Ion Product.*

SELECTIVE PRECIPITATION

The scheme of Qualitative Analysis is based on the *Principle of Selective Precipitation*. Let us consider a mixture containing Ag^+, Cu^{2+} and Fe^{3+}.

Step 1 : Add dil HCl to the solution of the mixture. Ag^+ will precipitate as AgCl. It is removed by filtration.

Step 2 : Pass H_2S through the filtrate. Cu^{2+} is precipitated as CuS and removed by filtration.

Step 3 : Add NH_4OH or NaOH solution to filtrate. Fe^{3+} is precipitated as $Fe(OH)_3$.

In this way ions are precipitated from solution one by one. *The precipitation of cations from solution one at a time is called Selective precipitation.*

Separation of the Basic Ions into Groups

Qualitative analysis of a mixture containing all the common cations involves first their separation into six major Groups based on solutions. Each group is then treated further to separate and identify the individual cations.

(I) Insoluble Chlorides

When excess dil HCl is added to a solution containing a mixture of common cations, only Ag^+, Pb^{2+} and Hg^{2+} will be precipitated as insoluble chlorides. The K_{sp} of these chlorides is very low and is easily exceeded. The K_{sp} value of other chlorides is high and they remain dissolved. For example, the K_{sp} of CuCl is 1.9×10^{-7} and, therefore, it will not be precipitated.

(II) Sulphides Insoluble in Acid Solution

After removing the insoluble chlorides in Group I, the solution is still acidic. When H_2S is passed through this solution, only the insoluble sulphides or Hg^{2+}, Cd^{2+}, Bi^{3+}, Cu^{2+} and Sn^{4+} will precipitate. It is so as the concentration of S^{2-} ion is relatively low because of the higher concentration of H^+ (Common Ion effect).

$$H_2S(g) \rightleftharpoons 2H^+ + S^{2-} \quad \text{(eqbm shifted to left)}$$

Other cations Mn^{2+}, Zn^{2+}, Ni^{2+}, CO^{2+} with higher value of K_{sp} of the respective sulphides will remain dissolved because the ionic product is less than K_{sp}.

(III) Insoluble Hydroxide in NH_4OH + NH_4Cl

To the filtrate from Group II is added NH_4Cl and then NH_4OH. NH_4OH is a weak base and the concentration of OH is low. NH_4Cl provides more NH_4^+ ions which shifts the equilibrium to the left (Common Ion effect).

$$NH_4OH \rightleftharpoons NH_4^+ + OH^-$$

Thus, some OH^- ions are converted to undissociated NH_4OH. So that the concentration of OH^- ions becomes extremely low. But the ion product, Q, is greater than K_{sp} values for the hydroxides of Al^{3+} Fe^{3+} and Cr^{3+} which precipitate out On the other hand, the ion product is lower than the K_{sp} values for $Zn(OH)_2$ $Mn(OH)_2$, and $Mg(OH)_2$ which remain dissolved.

	$Zn(OH)_2$	$MN(OH)_2$	$Mg(OH)_2$
K_{sp}	4.5×10^{-17}	4.6×10^{-14}	1.5×10^{-11}

(IV) Insoluble Sulphides in Basic Solution

The filtrate from Group III contains OH ions and is basic. H_2S is passed through it. In the presence of OH^- ions, the H^+ ions produced from H_2S form nonionised water.

$$H_2S \rightleftharpoons 2H^+ + S^{2-}$$

$$H^+ + OH \rightarrow H_2O \text{ (nonionised)}$$

Thus, the above equilibrium shifts to the light and the concentration of S2' ions increases. This increases the actual ion product for the reaction

$$\underset{\text{metal sulphide}}{MS(s)} \rightleftharpoons M^{2+} + S^{2-}$$

which becomes greater than the K_{sp} value for the insoluble sulphides of Mn^{2+}, Zn^{2+}, Ni^{2+} and CO^{2+}. Thus MnS, ZnS, NiS and CoS are therefore, precipitated.

(V) Insolute Carbonates

To the filtrate from Group (iv) is added excess of sodium carbonate solution. The excess CO_3^{2-} ions in solution will drive the equilibrium to the left for $BaCO_3$, $SrCO_3$, $MgCO_3$ and $CaCO_3$.

Thus

$$BaCO_3 \rightleftharpoons Ba^+ + CO_3^{2-} \text{ (excess)}$$

COLLOIDAL SYSTEMS

Thomas Graham (1861) studied the ability of dissolved substances to diffuse into water across a permeable membrane. He observed that crystalline substances such as sugar, urea, and sodium chloride passed through the membrane, while others like glue, gelatin and gum arable did not The former he called *crystalloids* and the latter *colloids* (Greek, *kolla* = glue; *eidos* = like). Graham thought that the difference in the behaviour of 'crystalloids' and 'colloids' was due to the particle size.

Later it was realised that *any substance, regardless of its nature, could be converted into a colloid by subdividing it into particles of colloidal size.*

In a true solution as sugar or salt in water, the solute particles are dispersed in the solvent as single molecules or ions. Thus the diameter of the dispersed particles ranges from 1 Å to 10 Å.

On the other hand, in a *suspension* as sand stirred into water, the dispersed particles are aggregates of millions of molecules. The diameter of these particles is of the order of 2000 Å or more.

The colloidal solutions or colloidal dispersions are intermediate between true solutions and suspensions. In other words, the diameter of the dispersed particles in a colloidal dispersion is more than that of the solute particles in a true solution and smaller than that of a suspension.

When the diameter of the particles of a substance dispersed in a solvent ranges from about 10 Å to 2,000 Å, the system is termed a colloidal solution, colloidal dispersion, or simply a colloid.

The material with particle size in the colloidal range is said to be in the colloidal state.

The colloidal particles are not necessarily corpuscular in shape. In fact, these may be rod-like, disc-like, thin films, or long filaments. For matter in the form of corpuscles, the diameter gives a measure of the particle size. However, in other cases one of the dimensions (length, width, and thickness) has to be in the colloidal range for the material to be classed as colloidal.

TYPICAL SOLS

Sols are colloidal systems in which a solid is dispersed in a liquid. These can be subdivided into two classes :

(a) Lyophilic sols (solvent-loving)

(b) Lyophobic sols (solvent-hating).

Lyophilic sols are those in which the dispersed phase exhibits a definite affinity for the medium or the solvent.

The examples of lyophilic sols are dispersions of starch, gum, and protein in water.

Lyophobic sols are those in which the dispersed phase has no attraction for the medium or the solvent.

The examples of lyophobic sols are dispersion of gold, iron (III) hydroxide and sulphur in water. The affinity or attraction of the sol particles for the medium, in a lyophilic sol, is due to hydrogen bonding with water. If the dispersed phase is a protein (as in egg) hydrogen bonding takes place between water molecules and the amino groups (—NH—, —NH_2) of the protein molecule. In a dispersion of starch in water, hydrogen bonding occurs between water molecules and the —OH groups of the starch molecule. *There are no similar forces of attraction when sulphur or gold is dispersed in water.*

Some features of lyophilic and lyophobic sols are listed below.

Ease of Preparation

Lyophilic sols can be obtained straightaway by mixing the material (starch, protein) with a suitable solvent. The giant molecules of the material are of colloidal size and these at once pass into the colloidal form on account of interaction with the solvent.

Lyophobic sols are not obtained by simply mixing the solid material with the solvent.

Charge on Particles

Particles of a hydrophilic sol may have little or no charge at all.

Particles of a hydrophobic sol carry positive or negative charge which gives them stability.

Solvation

Hydrophilic sol particles are generally solvated. That is, they are surrounded by an adsorbed layer of the dispersion medium which does not permit them to come together and coagulate. Hydration of gelatin is an example.

There is no solvation of the hydrophobic sol particles for want of interaction with the medium.

Viscosity

Lyophilic sols are viscous as the particle size increases due to solvation, and the proportion of free medium decreases. Warm solutions of the dispersed phase on cooling set to a gel *e.g.*, preparation of *table jelly*. Viscosity of hydrophobic sol is almost the same as of the dispersion medium itself.

Precipitation

Lyophilic sols are precipitated (or coagulated) only by high concentration of the electrolytes when the sol particles are desolvated.

Lyophobic sols are precipitated even by low concentration of electrolytes, the protective layer being absent.

Reversibility

The dispersed phase of lyophilic sols when separated by coagulation or by evaporation of the medium, can be reconverted into the colloidal form just on mixing with the dispersion medium. Therefore this type of sols are designated as *Reversible sols*.

- **Tyndall Effect**

Lyophilic Sols do not scatter light and show no Tyndall effect. These particles are large enough to exhibit Tyndall effect.

- **Migration in Electric Field**

Lyophilic sol particles (proteins) migrate to anode or cathode, or not at all, when placed in electric field. Lyophobic sol particles move either to anode or cathode, according to they carry positive charge or negative charge.

Differences between Lyophilic sols and Lyophobic sols

Lyophilic Sols	Lyophobic Sols
• Little or no charge on particles	• Particles carry positive or negative charge
• Reversible	• Irreversible
• Do not exhibit Tyndall effect	• Exhibit Tyndall effect
• Particles generally solvated	• No solvation of particles

PREPARATION OF LYOPHOBIC SOLS

Lyophobic sols can be prepared by two methods.

- Dispersion Method.
- Aggregation Method.

In Dispersion method, material in bulk is dispersed in another.

Mechanical Dispersion Using Colloid Mill

The solid along with the liquid dispersion medium is fed into a Colloid mill. The mill consists of two steel plates nearly touching each other and rotating in opposite directions with high speed. The solid particles are ground down to colloidal size and are then dispersed in the liquid to give the sol. 'Colloidal graphite' (a lubricant) and printing inks are made by this method. Recently, mercury sol has been prepared by disintegrating a layer of mercury into sol particles in water by means of ultrasonic vibrations.

Bredig's Arc Method

It is used for preparing hydrosols of metals *e.g.*, silver, gold and platinum. An arc is struck between the two metal electrodes held close together beneath *de-ionized water*. The water is kept cold by immersing the container in ice/water bath and a trace of alkali (KOH) is added.

The intense heat of the spark across the electrodes vaporises some of the metal and the vapour condenses under water. Thus the atoms of the metal present in the vapour aggregate to form colloidal particles in water. Since the metal has been ultimately converted into sol particles (via metal vapour), this method has been treated as of dispersion.

Non-metal sols can be made by suspending coarse particles of the substance in the dispersion medium and striking an arc between iron electrodes.

The more important methods of preparing hydrophic sols are as follows :

1. Double Decomposition
2. Preduction
3. Oxidation
4. Hydrolysis
5. Change of Solvent.

In the methods of preparation, the resulting sol frequently contains besides colloidal particles appreciable amounts of electrolytes. To obtain the pure sol, these electrolytes have to be removed. This purification of sols can be accomplished by three methods :

(a) Dialysis

(b) Electrodialysis

(c) Ultrafiltration

Dialysis

Animal membranes (bladder) or those made of parchment paper and cellophane sheet, have very fine pores. These pores permit ions (or small molecules) to pass through but not the large colloidal particles. When a sol containing dissolved ions (electrolyte) or molecules is placed in a bag of permeable membrane dipping in pure water, the ions diffuse through the membrane. By using a continuous flow of fresh water, the concentration of the electrolyte outside the membrane tends to be zero. Thus diffusion of the ions into pure water remains brisk all the time. In this way, practically all the electrolyte present in the sol can be removed easily.

KINETIC PROPERTIES OF SOLS

Brownian Movement

When a sol is examined with an ultramicroscope, the suspended particles are seen as shining specks of light By following an individual particle, it is observed that the particle is undergoing a constant rapid motion. It moves in a series of short straight-line paths in the medium, changing directions abruptly.

The continuous rapid zig-zag movement executed by a colloidal particle m the dispersion medium is called Brownian movement or motion.

This phenomenon is so named after Sir Robert Brown who discovered it in 1827.

Suspensions and true solutions do not exhibit Brownian movement.

Explanation of Brownian Movement

The explanation of Brownian movement was advanced by Albert Einstein around 1955 by mathematical considerations based on the kinetic molecular theory. According to him, at any instant a colloidal particle was being struck by several molecules of the dispersion medium.

The movement of the particle was caused by unequal number of molecules of the medium striking it from opposite directions. When more molecules struck the particle on one side than on another, the direction of movement changed. How a colloidal particle is knocked about in a zig-zag path by molecules of the dispersion medium. In a suspension, the suspended particles being very large, the probability of unequal bombardments diminishes. The force of the molecules hitting the particle on one side is cancelled by the force of collisions occurring on the other side. Hence they do not exhibit Brownian movement.

The phenomenon of Brownian movement is an excellent proof of the existence of molecules and their ceaseless motion in liquids. It also explains how the action of gravity, which would ordinarily cause the settling of colloidal particles, is counteracted. The constant pushing of the particles by the molecules of the dispersion medium has a stirring effect which does not permit the particles to settle.

Electrical properties of sols can be studied as follows.

The most important property of colloidal dispersions is that all the suspended particles possess either a positive or negative charge.

The mutual forces of repulsion between similarly charged particles prevent them from aggregating and settling under the action of gravity. This gives stability to the sol.

The sol particles acquire positive or negative charge by preferential adsorption of +ve or –ve ions from the dispersion medium. For example, a ferric hydroxide sol particles are positively charged because these adsorb Fe^{3+} ions from ferric chloride ($FeCl_3$) used in the preparation of the sol. Since the sol as a whole is neutral, the charge on the particle is counterbalanced by oppositely charged ions termed *counterions* (in this case Cl^-) furnished by the electrolyte in medium.

Electro-Osmosis

A sol is electrically neutral. Therefore the dispersion medium carries an equal but opposite charge to that of the dispersed particles. Thus, the medium will move in opposite direction to the dispersed phase under the influence of applied electric potential. When the dispersed phase is kept stationary, the medium is actually found to move to the electrode of opposite sign than its own.

The movement of the dispersion medium under the influence of applied potential is known as electro-osmosis.

Electro-osmosis is a direct consequence of the existence of zeta potential between the sol particles and the medium. When the applied pressure exceeds the zeta potential, the *diffuse layer* moves and causes electro-osmosis. The phenomenon of electro-osmosis can be demonstrated by using a U-tube in which a plug of wet clay (a negative colloid) is fixed.

The two limbs of the tube are filled with water to the same level. The platinum electrodes are immersed in water and potential applied across them. It will be observed that water level rises on the cathode side and falls on anode side. This movement of the medium towards the negative electrode, shows that the charge on the medium is +ve. Similarly, for a positively charged colloid electro-osmosis will take place in the reverse direction. Technically, the phenomenon has been applied in the removal of water from peat, in dewatering of moist clay and in drying dye pastes.

APPLICATIONS OF COLLOIDS

Colloids play an important role in our daily life and industry. A knowledge of colloid chemistry is essential to understand some of the various natural phenomena around us. Colloids make up some of our modem products. A few of the important applications of colloids are listed below.

Foods

Many of our foods are colloidal in nature. Milk is an emulsion of butterfat in water protected by a protein, casein. Salad dressing; gelatin deserts, fruit jellies and whipped cream are other.examples. Ice cream is a dispersion of ice in cream. Bread is a dispersion of air in baked dough.

Medicines

Colloidal medicines being finely divided, are more effective and are easily absorbed in our system. Halibut-liver oil and cod-liver oil that we take are, in fact, the emulsions of the respective oils in water. Many ointments for application to skin consist of physiologically active components dissolved in oil and made into an emulsion with water. Antibiotics such as penicillin and streptomycin are produced in colloidal form suitable for injections.

Non-Drip or Thixotropic Paints

All paints are colloidal dispersions of solid pigments in a liquid medium. The modem *nondrip* or *thixotropic paints* also contain long-chain polymers. At rest, the chains of molecules are coiled and entrap much dispersion medium. Thus the paint is a semi-solid gel structure. When shearing stress is applied with a paint brush, the coiled molecules straighten and the entrapped medium is released. As soon as the brush is removed, the liquid paint reverts to the semi-solid form. This renders the paint 'non-drip'.

Electrical Precipitation of Smoke

The smoke coming from industrial plants is a colloidal dispersion of solid particles (carbon, arsenic compounds, cement dust) in air. It is a nuisance and pollutes the atmosphere. Therefore, before allowing the smoke to escape into air, it is treated by *Cottrell Precipitator.* The smoke is led past a series of sharp points charged to a high potential (20,000 to 70,000V).

The points discharge high velocity electrons that ionise molecules in air. Smoke particles adsorb these positive ions and become charged. The charged particles are attracted to the oppositely charged electrodes and get precipitated. The gases that leave the Cottrell precipitator are thus freed from smoke. In addition, valuable materials may be recovered from the precipitated smoke. For example, arsenic oxide is mainly recovered from the smelter smoke by this method.

Clarification of Municipal Water

The municipal water obtained from natural sources often contains colloidal particles. The process of coagulation is used to remove these. The sol particles carry a negative charge. When aluminium sulphate (*alum*) is added to water, a gelatinous precipitate of hydrated aluminium hydroxide (*floc*) is formed,

$$Al_3^+ + 3H_2O \rightarrow Al(OH)_3 + 3H^+$$

$$Al(OH)_3 + 4H_2O \rightarrow Al(OH)_3(H_2O)_4^+.$$

The positively charged *floc* attracts to it negative sol particles which are coagulated. The *floc* along with the suspended matter comes down, leaving the water clear.

Formation of Delta

The river water contains colloidal particles of sand and clay which carry negative charge. The sea water, on the other hand, contains positive ions such as Na^+, Mg^{2+}, Ca^{2+}. As the river water meets sea water, these ions discharge the sand or clay particles which are precipitated as *delta.*

Artificial Kidney Machine

The human kidneys purify the blood by dialysis through natural membranes. The toxic waste products such as urea and uric acid pass through the membranes, while colloidal-sized particles of blood proteins (haemoglobin) are retained. Kidney failure, therefore, leads to death due to accumulation of poisonous waste products in blood. Now-a-days, the patient's blood can be cleansed by shunting it into an 'artificial kidney machine'. Here the impure blood is made to pass through a series of *cellophane tubes* surrounded by a washing solution in water. The toxic waste chemicals (urea, uric acid) diffuse across the tube walls into the washing solution. The purified blood is returned to the patient. The use of artificial kidney machine saves the life of thousands of persons each year.

Adsorption Indicators

These indicators function by preferential adsorption of ions onto sol particles. fluorescein (Na^+Fl^-) is an example of adsorption indicator which is used for the titration of sodium chloride solution against silver nitrate solution.

When silver nitrate solution is run into a solution of sodium chloride containing a little fluorescein, a white precipitate of silver chloride in first formed. At the end-point, the white precipitate turns sharply pink.

EMULSIONS

Emulsions are liquid-colloidal systems, or, an emulsion may be defined as a dispersion of finely divided liquid droplets in another liquid.

Generally, one of the two liquids is *water* and the other, which is immiscible with water, is designated as *oil*. Either liquid can constitute the dispersed phase.

Types of Emulsions

There are two types of Emulsions.

(a) Oil-in-Water type (O/W type);

(b) Water-in-Oil type (W/O type)

Examples of Emulsions

(1) Milk is an emulsion of O/W type. Tiny droplets of liquid fat are dispersed in water.

(2) Stiff greases are emulsions of W/O type, water being dispersed in lubricating oil.

Preparation of Emulsions

The dispersal of a liquid in the form of an emulsion is called *emulsification*. This can be done by agitating a small proportion of one liquid with the bulk of the other. It is better accomplished by passing a mixture of the two liquids through a colloid mill known as *homogenizer*.

The emulsions obtained simply by shaking the two liquids are unstable. The droplets of the dispersed phase coalesce and form a separate layer. To have a stable emulsion, small amount of a third substance called the *Emulsifier* or *Emulsifying agent* is added during the preparation. This is usually a soap, synthetic detergent, or a hydrophilic colloid.

Role of Emulsifier

The emulsifier concentrates at the interface and reduces surface tension on the side of one liquid which rolls into droplets. Soap, for example, is made of a long hydrocarbon tail (oil soluble) with a polar head $—COO^{-}Na^{+}$ (water soluble).

In O/W type emulsion the tail is pegged into the oil droplet, while the head extends into water. Thus, the soap acts as go-between and the emulsified droplets are not allowed to coalesce.

4

Specific Conductance

INTRODUCTION

The phenomenon of decomposition of an electrolyte by passing electric current through its solution is termed Electrolysis (*l*yo = breaking). The process of electrolysis is carried in an apparatus called the *Electrolytic cell*. The cell contains water-solution of an electrolyte in which dip two metallic rods, the *Electrodes*.

These rods are connected to the two terminals of a battery (source of electricity). The electrode connected to the positive terminal of the battery attracts the negative ions (anions) and is called *anode*. The other electrode connected to the negative end of the battery attracts the positive ions (cations) and is called *cathode*.

Water-soluble substances are distinguished as *electrolytes* or *nonelectrolytes*.

Electrolytes are electrovalent substances that form ions in solution which conduct an electric current.

Sodium chloride, copper (II) sulphate and potassium nitrate are examples of electrolytes

Non-electrolytes, on the other hand, are covalent substances which furnish neutral molecules in solution. Their water-solutions do not conduct an electric current. Sugar, alcohol and glycerol are typical nonelectrolytes.

How the electrolysis actually takes place, is illustrated. The cations migrate to the cathode and form a neutral atom by accepting electrons from it. The anions migrate to the anode and yield a neutral particle by transfer of electrons to it. As a result of the loss of electrons by anions and gain of electrons by cations at their respective electrodes, chemical reaction takes place.

Let us consider the electrolysis of hydrochloric acid as an example. In solution, HCl is ionised,

$$HCl \rightarrow H^+ + Cl^-$$

In the electrolytic cell Cl^- ions will move toward the anode and H^+ ions will move toward the cathode. At the electrodes, the following reactions will take place.

At cathode :

$$H^+ + e^- \rightarrow H \qquad \text{(Reduction)}$$

Each hydrogen ion picks up an electron from the cathode to become a hydrogen atom pairs of hydrogen atoms then unite to form molecules of hydrogen gas H_2.

At Anode:

$$Cl^- \rightarrow Cl + e^- \qquad \text{(Oxidation)}$$

After the chloride ion loses its electron to the anode, pairs of chlorine atoms unite to form chlorine gas, Cl_2.

The net effect of the process is the decomposition of HCl into hydrogen and chlorine gases. The overall reaction is:

$$2HCl \rightarrow H_2 + Cl, \qquad \text{(Decomposition)}$$

The electrical conductivity of solutions is reported as the specific conductance.

The fact that aqueous solutions of certain materials, called electrolytes, conduct an electric current provides the most direct evidence for the idea that ions capable of independent motion are present. More detailed studies of the electrical conductivity of such solutions provide information on the number and independence of these ions.

Measurements of the conductivity of aqueous solutions are made with a conductivity cell and an electric circuit like that when an alternating current is used to prevent buildup of charges of opposite sign near the two electrode surfaces, so that there is little electric resistance at the metal-solution interface, the conductivity cell obeys Ohm's law: The current flowing through the cell is proportional to the voltage across the cell. It is therefore possible to assign a resistance of so many ohms to such a cell, just as one assigns a resistance to a metallic conductor.

It is more convenient to focus on the *conductance* of an electrolytic solution rather than on its *resistance*. These quantities are reciprocally related, and the conductance L is calculated from the measured resistance as

$$L = \frac{1}{R} \qquad ...(1)$$

If R is the resistance in ohms, symbol Ω, then L has the units of Ω^{-1}. As for metallic conductors, the resistance and therefore the conductance depend on the cross-section area A and the length I of the region between the electrodes. As for a metallic conductor,

$$R = \rho\frac{l}{A} \qquad ...(2)$$

where p is the *specific resistance* and is the proportionality factor that corresponds to the resistance of a cell of unit cross-section area and unit length. One can also write

$$L = \kappa\frac{A}{l} \qquad ...(3)$$

where K, the *specific conductance*, can be thought of as the conductance of a cube of the solution of electrolyte of unit dimensions.

The specific conductance can, in principle, be obtained from the measured value of R, which gives L = l/R, and of l and A of the cell. In practice, it is more convenient to deduce I and A, or rather the cell constant I/A, from a measurement of L when the cell is filled with a solution of known specific conductance. Once this geometric factor has been obtained for a cell, it can be used to deduce K for an unknown solution from a measured value of L and Eq. (3.)

The cell constant is almost always determined by using a solution of KCl. Specific conductances of these reference solutions have been determined by measurements with rather elaborately designed electrodes which avoid the uncertainty in the effective current-carrying cross section that exists in ordinary cells. Since the strong temperature dependence is characteristic of all conductance results, it is necessary to make measurements of conductances in well-thermostated cells. Although, the specific conductance is a measure of the ease with which a current flows through a unit cube of solution, it is not a convenient quantity for the discussion of the conduction process of solutions of electrolytes.

TYPES OF ELECTROLYTES

Electrolytes may be divided into two classes :

(a) Strong Electrolytes

(b) Weak Electrolytes.

Strong Electrolytes

A strong electrolyte is a substance that gives a solution in which almost all the molecules are ionised. The solution itself is called a strong electrolytic solution. Such solutions are good conductors of electricity and have a high value of equivalent conductance even at low concentrations. The strong electrolytes are :

(1) *The strong acids e.g.*, HCl, H_2SO_4, HNO_3, $HClO_4$, HBr and HI.

(2) *The strong bases e.g.*, NaOH, KOH, $Ca(OH)_2$, $Mg(OH)_2$, etc.

(3) *The salts* : Practically all salts (NaCl, KCl, etc.) are strong electrolytes.

Weak Electrolytes

A weak electrolyte is a substance that gives a solution in which only a small proportion of the solute molecules are ionised. Such a solution is called a *weak electrolytic solution*, that has low value of equivalent conductance. The weak electrolytes are:

(1) *The Weak Acids* : All organic acids such as acetic acid, oxalic acid, sulphurous acid (H_2SO_3) are examples of weak electrolytes.

(2) *The Weak Bases* : Most organic bases *e.g.*, alkyl amines ($C_2H_5NH_2$) are weak electrolytes.

(3) *Salts* : A few salts such as mercury (II) chloride and lead (II) acetate are weak electrolytes.

Measurement of Electrolytic Conductance

We know that conductance is the reciprocal of resistance. Therefore it can be determined by measuring the resistance of the electrolytic solution. This can be done in the laboratory with the help of a *wheatstone bridge*. The solution whose conductance is to be determined is placed in a special type of cell known as the *conductance cell*. A simple type of conductance cell used in the laboratory. The electrodes fitted in the cell are made of platinum plates coated with platinum black. These are welded to platinum wires fused in two thin glass tubes. The contact with copper wires of the circuit is made by dipping them in mercury contained in the tubes.

TRANSFERENCE NUMBERS

The transference number gives the fraction of the current carried through a solution by an ion of a particular type.

Now that the general features of electrode processes have been mentioned, the details of the passage of the electric current through the body of the solution can be investigated. Since the flow of either the positive or the negative ions, or both, might be responsible for conduction processes, our first goal is the determination of the fraction of the current carried by each type of ion of a given electrolyte. For this purpose the transference numbers t_+ and t_- are introduced according to the definitions

t_+ = fraction of current carried by cation

t_- = fraction of current carried by anion.

The fractional character of these transference numbers implies

$$t_+ + t_- = 1.$$

In metal conductors, *e.g.*, a copper wire, all the current is carried by the electrons, and for such conductors one could write $t_- = 1$ and $t_+ = 0$. For solutions of electrolytes it is clearly difficult to anticipate what fraction of the current is carried past some position in the electrolyte by the cations and what fraction by the anions.

One method, known as the *Hittorf Method*, for measuring transference numbers is now illustrated by two examples. A schematic diagram of a cell marked off into three compartments. In practice, a cell of the type can be used, and the three compartments that can be drained off correspond to those marked off by the cross-section lines. The following treatment will show that transference numbers can be deduced from the analysis for the amount of electrolyte in the separate compartments following passage of a measured amount of current through the cell.

CALCULATION OF IONIC MOBILITIES

Values can now be obtained for the contributions the individual ions of an electrolyte make to the molar conductance. The empirical law of Kohlrausch implies that at infinite dilution the molar conductance can be interpreted in terms of such ionic contributions and that the contributions of an ion are independent of the other ion of the electrolyte. At infinite dilution,

$$\Lambda^\circ = v_+ \lambda_+^\circ + v_- \lambda_-^\circ$$

where λ_+° and λ_-° are the *molar ionic conductances at infinite dilution.* Since the transference numbers give the fraction of the total current carried by each ion, *i.e.*, the fraction of the total conductance that each ion contributes, we can write

$$\nu_+ \lambda_+^\circ = t_+^\circ \Lambda^\circ \quad \text{and} \quad \nu_- \lambda_-^\circ = t_-^\circ \Lambda^\circ$$

where t_+° and t_-° are the transference numbers extrapolated to infinite dilution. Some values for these limiting molar ionic conductances. An immediate use of such tabulated values is the calculation of the limiting molar conductance of a weak electrolyte without the addition and subtraction procedures.

The data for the molar conductance and the transference numbers at any concentration can be used to obtain values of λ_+ and λ_-. At concentrations other than that of infinite dilution, however, the law of independent migration of the ions fails, and the conductance is really a property of the electrolyte rather than of the individual ions of the electrolyte. This means that an ionic conductance calculated for a Cl^- ion, for example in a 1 M HCl solution, will be different from that deduced for the Cl^- ion in a 1 M NaCl solution. The concept of ionic conductances is really valuable, therefore, only at infinite dilutions.

Instead of trying to understand the different current-carrying properties of the ions in terms of ionic conductances, we can obtain an even more fundamental ionic property, the velocity with which the ions travel through the solution under the influence of the applied electric potential.

IONIC CONDUCTANCE

Arming any ion in solution there exists an ion *atmosphere* containing a preponderance of the ions with charges opposite that of the central ion. The ionic distribution can be looked on as resembling, for example, that in an expanded and loosely held NaCl crystal. The overall arrangement places each ion among nearest neighbours of the opposite charge. This ion atmosphere around each ion is, of course, better formed at higher concentrations.

The application of an electric field, as in a conductance experiment, results initially in the movement of the central ion away from the center of the oppositely charged sphere. The distorted ion atmosphere tends to oppose the applied field, and this decreases the current produced by a given applied electric field. Since the ion atmosphere is more important at higher concentrations, this decrease becomes more important at higher concentrations. Evaluation of this factor by P. Debye and H. Hückel showed that it contributes to the $\sqrt{c}$ dependence of the equivalent conductance. This effect is further enhanced by the tendency of the oppositely charged ions that predominate in the ion atmosphere to move in the opposite direction.

This ionic-atmosphere drag depends on the fact that the atmosphere does not instantaneously adjust itself to the new positions of the central ion. One says that the ionic atmosphere has a *relaxation time*; *i.e.*, when a stress is applied, it takes a finite time for the atmosphere to relax, or to be reestablished. The mechanism by which this occurs, as the central ion moves, is better thought of as a process of building new ions on the front of the atmosphere and dropping some off the back than of maintaining the same set of ions, which move to keep up with the central ion.

The second factor that acts to decrease the conductance at higher concentrations is an enhanced frictional drag that sets in. When an electric field is applied, the ions set off to the oppositely charged electrodes. Each ion moves with a velocity that depends on a balance between the electric force and the viscous drag.

The average velocity and therefore the current are concentration-dependent because the ions can be thought of as carrying along with them their many solvating molecules, and at higher concentrations an ion seems to swim against the current produced by the oppositely charged, solvated ions moving in the opposite direction.

The conductivity theory of Debye and Hückel has been used to draw the slopes of the straight dashed lines. It is apparent that the effects considered by the theory are adequate to explain the conductance behaviour of strong electrolytes up to concentrations of about 0.01 M.

Thus the Arrhenius assumption that a decrease in conductance must be interpreted as a decreased number of conducting particles cannot be maintained. The relation $\alpha = \Lambda/\Lambda^\circ$ can be applied only when the Debye-Hückel effects are not appreciable or have been corrected for. For solutions with low ionic concentrations these effects will be small. For weak electrolytes, therefore, one can still rely on Λ/Λ° to give a value that can be interpreted primarily in terms of a degree of dissociation.

IONS OF A WEAK ELECTROLYTE

Coefficients of the ions of a weak electrolyte can be deduced from dissociation equilibria.

Studies of chemical equilibria can be used to deduce thermodynamic properties of non ideal systems. Acid-base equilibria provide many illustrations. The traditional example is the equilibrium set up by the dissociation of acetic acid, CH_3COOH, abbreviated HAc.

Introduction of the mean activity coefficient,

$$r \pm \sqrt{r+r-}\ R_{H}^{+} r_{Ac} - \frac{[H^{+}][Ac^{-}]}{[HAc]} \quad ...(1)$$

the rearrangement of the equation gives,

$$\log \frac{[H^{+}][Ac^{-}]}{[HAc]} = \log k_{th} - 2 \log 4_{\pm} \quad ...(2)$$

The concentration expression is that for the equilibrium constant interms of concentrations and can be written interms of the degree of dissociation α.

Then with $[H^{+}] = [Ac^{-}] = c\alpha$ and $[HAc] = c(1 - \alpha)$. Eq. (2) becomes

$$\log \frac{c\alpha^{2}}{1 - \alpha} = \log K_{th} - 2 \log \gamma_{\pm}$$

For solutions that are very dilute in ions, the Arrhenius expression $\alpha = \Lambda/\Lambda^{\circ}$ can be used to obtain the degree of dissociation from the conductivity. The left side can be determined for various acetic acid concentrations from conductivity measurements. The right side consists of a constant term $\log K_{th}$ and a term which, in view of the discussion can be expected to be a function of μ, the ionic strength. If the solution contains only the H^{+} and Ac^{-} ions from the dissociation of HAc, the ionic strength is given by

$$\mu = 1/2\ [(c\alpha)(1)^{2} + (c\alpha)(-1)^{2}] = c\alpha$$

The theory of Debye and Hückel, developed suggests that in the low-ionic-concentration range the relation between $\gamma_{\pm}$ and μ is such that $\log \gamma_{\pm}$ is proportional to $\sqrt{\mu}$. It might therefore be informative to plot the left side of against $\sqrt{c\alpha}$,. At low concentrations the points do seem to fall along a straight line, in agreement with the prediction of the Debye-Hückel theory. (The theory predicts, furthermore, that the last term should be $-2 \log \gamma_{\pm} = +(2)(0.509)\sqrt{c\alpha} = \sqrt{c\alpha}\ 1.018$. The line drawn with slope +1.018, fits the data satisfactorily.)

Extrapolation to zero ionic strength, where $\gamma_{\pm} = 1$ and $\log \gamma_{\pm} = 0$, gives

$$\log K_{th} = -4.7565$$

and $\quad K_{th} = 1.752 \times 10.5$

Equation can be used, with the known value of K_{th}, to obtain values for the mean-activity coefficient. Rearrangement gives

$$\log \gamma_{\pm} = \frac{1}{2} \log K_{th} - \frac{1}{2} \log \frac{c\alpha^2}{1-\alpha} = -2.3782 - \frac{1}{2} \log \frac{c\alpha^2}{1-\alpha}$$

From this equation we can determine $\gamma_{\pm}$ for the H^+ and Ac^- ions of acetic acid solutions,

For example, at an acetic acid concentration of 0.01 mol L^{-1} the degree of dissociation $\alpha = \Lambda/\Lambda°$ from the conductance data of we obtain

$$\log \gamma_{\pm} = -2.3782 + 1/2(4.7409)$$

and $\gamma_{\pm} = 0.982.$

Notice, finally, that at [HAc] = 0.01, $[H^+][Ac^-]/[HAc] = K_{th}/\gamma_{\pm}^2$ 1.82 × 10.5, appreciably different from $K_{th} = 1.752 \times 10^{-5}$.

CHEMICAL BONDING

An understanding of atomic structure and of the methods of wave mechanics provides the means for tackling one of the fundamental questions of chemistry: What binds atoms together into molecules? This question has existed since the beginnings of chemistry, and a clear answer would be a culmination of much of the theoretical work of chemistry.

The quantum-mechanical approach to covalent bonding will be introduced by considering the bonding in the hydrogen-molecule ion and the hydrogen molecule. From these relatively simple systems it will be possible to extend the theory of chemical bonding in a semiquantitative and semiempirical way. It is this extension which has become a basic and necessary part of the approaches and language in all branches of chemistry.

Molecules of chemical substances are made of two or more atoms joined together by some force, acting between them. This force which results from the interaction between the various atoms that go to form a stable molecule, is referred to as *Chemical Bond*. A chemical bond is defined as a force that acts between two or more atoms to hold them together as a stable molecule.

As we will study later, there arc three different types of bonds recognised by chemists:

(1) Ionic or Electrovalent Bond

(2) Covalent Bond

(3) Coordinate Covalent Bond

In order to understand about chemical bonding it is very much needed to study about valence electrons.

The electrons in the outer energy level of an atom are the ones that can take part in chemical bonding. These electrons are, therefore, referred to as the valence electrons.

The electronic configuration of Na is 2, 8, 1 and that of Cl is 2, 8, 7. Thus sodium has one valence electron and chlorine 7. It is important to remember that for an A group element of the periodic table (H, O, K, F, Al etc.) the group number is equal to the number of valence electrons,

Bonding and Non-bonding Electrons

The valence electrons actually involved in bond formation are called bonding electrons. The remaining valence electrons still available for bond formation are referred to as non-bonding electrons.

IONIC BOND

This is the type of bond which is established by transfer of an electron from one atom to another.

The following factors influence very much in the formation of Ionic Bond.

Number of Valence Electrons

The atom A should possess 1, 2 or 3 valence electrons, while the atom B should have 5, 6 or 7 valence electrons. The elements of group IA, IIA and IIIA satisfy this condition for atom A and those of groups VA, VIA, and VIIA satisfy this condition for atom B.

Net Lowering of Energy

To form a stable ionic compound, there must be a net lowering of the energy. In other words energy must be released as a result of the electron transfer and formation of ionic compound by the following steps:

(a) The removal of electron from atom A ($A - e^- \rightarrow A^+$) requires input of energy, which is the ionization energy (IE). It should be low.

(b) The addition of an electron to B ($B + e^- \rightarrow B^-$) releases energy, which is the electron affinity of B(EA). It should be high.

(c) The electrostatic attraction between A^+ and B^- in the solid compound releases energy, which is the electrical energy. It should also be high.

If the energy released in steps (b) and (c) is greater than the energy consumed in step (a), the overall process of electron transfer and formation of ionic compound results in a net release of energy. Therefore, ionisation of A will occur and the ionic bond will be formed.

Ionisation Energy

The ionisation energy of the metal atom which loses electron(s) should be low so that the formation of +vely charged ion is easier. *Lower the ionisation energy greater will he the tendency of the metal atom to change into cation and hence greater will be the ease of formation of ionic bond.* That is why *alkali metals and alkaline earth metals form ionic bonds easily.* Out of these two, alkali metals form ionic bonds easily as compared to alkaline earth metals. *In a group the ionisation energy decreases as we move down the group and therefore, the tendency to form ionic bond increases in a group downward.* Due to this reason Cs is the most electropositive atom among the alkali metals.

Electron Affinity

The atom which accepts the electron and changes into anion should have high electron affinity. *Higher the electron affinity more is the energy released and stable will be the onion formed.* The A elements of group VIA and VIIA have, in general, higher electron affinity and have high tendency to form ionic bonds. Out of these two, *the elements of group VIIA (halogens) are more prone to the formation of ionic bond than the elements of group* VIA. In moving down a group the electron affinity decreases and, therefore, the tendency to form ionic bond also decreases.

Lattice Energy

After the formation of cations and anions separately, they combine to form ionic compound. In this process, energy is released. It is called *Lattice Energy*. It may be defined as "the amount of energy released when one mole of an ionic compound is formed from its cations and anions."

Greater the lattice energy, greater the strength of ionic bond. The value of lattice energy depends upon the following two factors:

Size of the Ions

In order to have the greater force of attraction between the cations and anions their size should be small as the force of attraction is inversely proportional to the square of the distance between them.

Charge on Ions

Greater the charge on ions greater will be the force of attraction between them and, therefore, greater will be the strength of the ionic bond.

Now let us study about the characteristics of Ionic compounds.

On account of strong electrostratic forces between the opposite ions, these ions are locked in their allotted positions in the crystal lattice. Since they lack the freedom of movement characteristic of the liquid state, they are solids at room temperature.

The ionic compounds are made of (+) and (–) ions held by electrostatic forces in a crystal lattice. Each ion is surrounded by the opposite ions in alternate positions in a definite order in all directions. This explains the common properties of ionic compounds.

Solids at Room Temperature

Ionic compounds have high melting points (or boiling points). Since the (+) and (–) ions are tightly held in their positions in the lattice, only at high temperature do the ions acquire sufficient kinetic energy to overcome their attractive forces and attain the freedom of movement as in a liquid. Thus ionic compounds need heating to high temperatures before melting.

Hard and Brittle

The crystals of ionic substances re hard and brittle. Their hardness is due to the strong electrostatic forces which hold each ion in its allotted position.

These crystals are made of layers of (+) and (–) ions in alternate positions so that the opposite ions in the various parallel layers lie over each other. When external force is applied to a layer of ions with respect to the next, even a slight shift brings the like ions in front of each other. The (+) and (–) ions in the two layers thus repel each other and fall apart. The crystal cleaves here.

Soluble in Water

When a crystal of an ionic substance is placed in water, the polar water molecules detach the (+) and (–) ions from the crystal lattice by their electrostatic pull. These ions then get surrounded by water molecules and can lead an independent existence and are thus dissolved in water. By the same reason, non-polar solvents like benzene (C_6H_6) and hexane

(C_6H_{14}) will not dissolve ionic compounds. Solid ionic compounds are poor conductors of electricity because the ions are fixed rigidly in their positions. In the molten stale and in water solutions, ions are rendered free to move about. Thus molten ionic compounds or their aqueous solutions conduct a current when placed in an electrolytic cell.

Do not Exhibit Isomerism

The ionic bond involving electrostatic lines of force between opposite ions, is non-rigid and non-directional. The ionic compounds, therefore, are incapable of exhibiting stereoisomerism.

Ionic Reactions are Fast

Ionic compounds give reactions between ions and these are very fast

COVALENT BOND

A shared pair of electrons constitutes a covalent bond or Electron-pair bond. The compounds containing a covalent bond are called covalent compounds.

The conditions which are required for the formation of covalent bond are as follows :

Number of Valence Electrons

Each of the atoms A and B should have 5, 6 or 7 valence electrons so that both achieve the-stable octet by sharing 3, 2 or 1 electron-pair. H has one electron in the valence shell and attains duplet. The non-metals of groups VA, VIA and VIIA respectively satisfy this condition.

Equal Electronegativity

The atom A will not transfer electron/s to B if both have equal electronegativity, and hence electron sharing will take place. This can be strictly possible only if both the atoms are of the same element.

In many molecules, we find that in order to satisfy the octet, it becomes necessary for two atoms to share two or three pairs of electrons between the same two atoms. The sharing of two pairs of electrons is known as a Double bond and the sharing of three pairs of electrons a *Triple bond*. Let us consider some examples of compounds containing these multiple covalent bonds in their molecules.

Let us study about the characteristics of covalent compounds.

On application of heat, the molecules are readily pulled out and these then acquire kinetic energy for free movement as in a liquid. For the same reason, the liquid molecules are easily obtained in the gaseous form which explains low boiling points of covalent liquids.

Neither Hard nor Brittle

While the ionic compounds are hard and brittle, covalent compounds are neither hard nor brittle. There are weak forces holding the molecules in the solid crystal lattice. A molecular layer in the crystal easily slips relative to other adjacent layers and there are no 'forces of repulsion' like those in ionic compounds. Thus, the crystals are easily broken and there is no sharp cleavage between the layers on application of external force,

Soluble in Organic Solvents

In general, covalent compounds dissolve readily in nonpolar organic solvents (benzene, ether). The kinetic energy of the solvent molecules easily overcomes the weak intermolecular, forces.

Covalent compounds are insoluble in water. Some of them (alcohols, amines) dissolve in water due to hydrogen-bonding.

Non-Conductors of Electricity

Since there are no (+) or (–) ions in covalent molecules, the covalent compounds in the molten or solution form are incapable of conducting electricity.

Exhibit Isomerism

Covalent bonds are rigid and directional, the atoms being held together by shared electron pair and not by electrical lines of force. This affords opportunity for various spatial arrangements and covalent compounds exhibit stereoisomerism.

Molecular Reactions

The covalent compounds give reactions where the molecule as a whole undergoes a change. Since there are no strong electrical forces to speed up the reaction between molecules, these reactions are slow.

In a normal covalent bond, each of the two bonded atoms contributes one electron to make the shared pair. In some cases, a covalent bond is formed when both the electrons are supplied entirely by one atom. Such a bond is called coordinate covalent or dative bond. It may be defined as: a covalent bond in which both electrons of the shared pair come from one of the two atoms (or ions).

The compounds containing a coordinate bond are called *coordinate compounds*.

If an atom A has an unshared pair of electrons (tone pair) and another atom B is short of two electrons than the stable number, a coordinate bond is formed. A donates the lone pair to B which accepts it Thus both A and B achieve the stable 2 or 8 electrons, the lone pair being held in common.

The atom A which donates the lone pair is called the *donor*, while B which accepts it the acceptor. The bond thus established is indicated by an arrow pointing from A to B. Although the arrow head indicates the origin of the electrons, once the coordinate bond is formed it is in no way different from an ordinary covalent bond.

In the H_2 or Cl_2 molecule, the two electrons constituting the covalent bond are equally shared by the two identical nuclei. Due to even distribution of (+) and (–) charge, the two bonded atoms remain electrically neutral. Such a bond is called nonpolar covalent bond. However, when two different atoms are joined by a covalent bond as in HCl, the electron pair is not shared equally.

A covalent bond in which electrons are shared unequally and the bonded atoms acquire a partial positive and negative charge, is called a polar covalent bond.

A molecule having partial positive and negative charge separated by a distance is commonly referred to as a Dipole (two poles). The dipole of a bond is indicated by an arrow from positive to negative end with a crossed tail as shown above in HCl molecule.

Since two atoms of different elements do not have exactly the same attraction for electrons in a bond, all bonds between unlike atoms are polar to some extent.

The amount of polarity of a bond is determined by the difference of electronegativity (or tendency to attract electrons) of the two bonded atoms. The greater the difference of electronegativity between two atoms, greater the polarity.

Now let us study some of the differences between Ionic and covalent bonds.

Ionic Bond	Covalent Bond
1. Formed by transfer of electrons from a metal to a non-metal atom.	1. Formed by sharing of electrons between non-metal atoms.
2. Consists of electrostatic force between (+) and (–) ions.	2. Consists of a shared pair of electrons between atoms.
3. Non-rigid and non-directional; cannot cause isomerism.	3. Rigid and directional; causes stereoisomerism.
Properties of Compounds	
1. Solids at room temperature.	1. Gases, liquids or soft solids.
2. High melting and boiling points.	2. Low melting and boiling points.
3. Hard and brittle.	3. Soft, much readily broken.
4. Soluble in water but insoluble in organic solvents.	4. Insoluble in water but soluble in organic solvents.
5. Conductors of electricity	5. Non-conductors of electricity.
6. Undergo ionic reactions which are fast.	6. Undergo molecular reactions which are slow.

HYDROGEN BONDING

The electrostatic attraction between an H atom covalently bonded to a highly electronegative atom X and a lone pair of electrons on X in another molecule, is called Hydrogen Bonding.

Hydrogen bond is represented by a dashed or dotted line.

It may be noted that:

(1) *Only O, N and F which have very high electronegativity and small atomic size, are capable of forming hydrogen bonds.*

(2) *Hydrogen bond is longer and much weaker than a normal covalent bond.* Hydrogen bond energy is less than 10 kcal/mole, while that of covalent bond is about 120 kcal/mole.

(3) Hydrogen bonding results in long chains or clusters of a large number of 'associated' molecules like many tiny magnets.

(4) *Like a covalent bond, hydrogen bond has a preferred bonding direction.* This is attributed to the fact that hydrogen bonding occurs through p orbitals which contain the lone pair of electrons on X atom. This implies that the atoms X–H...X will be in a straight line.

The necessary conditions for the formation of hydrogen bonding are

(i) *High electronegativity of atom bonded to hydrogen*

The molecule must contain an atom of high electronegativity such as F, O or N bonded to hydrogen atom by a covalent bond. The examples are HF, H_2O and NH_3.

(ii) *Small size of electronegative atom*

The electronegative atom attached to H-atom by a covalent bond should be quite small. Smaller the size of the atom, greater will be the attraction for the bonded electron pair. In other words, the polarity of the bond between H atom and electronegative atom should be high. This results in the formation of stronger hydrogen bonding. For example. N and Cl both have 3.0 electronegativity. But hydrogen bonding is effective in NH_3 in comparison to that in HCl It is due to smaller size of N atom than Cl atom.

When hydrogen bonding occurs between different molecules of the same compound as in HF, H_2O and NH_3 it is called *Intermolecular hydrogen bonding*. If the hydrogen bonding takes place within single molecule as in 2-nitrophenol, it is referred to as *Intramolecular hydrogen bonding*.

Types of Hydrogen—Bonding

Hydrogen bonding is of two types :

(i) Intermolecular Hydrogen bonding

(ii) Intramolecular Hydrogen bonding

(i) Intermolecular Hydrogen Bonding

The compounds in which molecules are joined to one another by hydrogen bonds, have usually, high boiling and melting points. This is because here relatively more energy is required to separate the molecules as they enter the gaseous state or the liquid state. Thus the hydrides of fluorine (HF), oxygen (H_2O) and nitrogen (NH_3) have abnormally high boiling and melting points compared to other hydrides of the same group which form no hydrogen bonds. The boiling points and melting points of the hydrides of VIA group elements plotted against molecular weights.

It will be noticed that there is a trend of decrease of boiling and melting points with decrease of molecular weight from H_2Te to H_2S. But there is a sharp increase in case of water (H_2O), although it has the smallest molecular weight. The reason is that the molecules of water are 'associated' by hydrogen bonds between them, while H_2Te, H_2Se and H_2S exist as single molecules since they are incapable of forming hydrogen bonds.

MOLECULAR ORBITALS

In an alternative description of bonding in polyatomic molecules atomic orbitals are combined to form molecular orbitals. Now let us study about the symmetries of the molecular orbitals of the H_2O molecule.

Let us consider the H_2O molecule with C_{2v} symmetry. The valence-shell 2s atomic orbital of the oxygen atom of the H_2O molecule is completely symmetric and, this is indicated by labelling this atomic central-atom orbital as Cl.

The symmetries of the P_x, P_y, and p, oxygen-atom orbitals are given by the locations of x, y, and z in the last column of the C_{2v} point-group table. The symmetries of the oxygen-atom p orbitals are indicated by a_1, b_1, and b_2.

The symmetries of attached-atom orbitals, or of combinations of these orbitals. The symmetries of the sum and difference of the two hydrogen atoms gives them the designations a_1 and b_1.

Central-atom and attached-atom orbitals with the same symmetry can be combined to give bonding and antibonding molecular orbitals with the same symmetry. For the water-molecule example, bonding and antibonding molecular orbitals with a_1 and b_1 symmetries are formed. The oxygen-atom b_2, orbital is not involved with the attached-atom orbitals, and its description in the molecule is unchanged from what it was in the atom.

The detailed shape and the relative energies of the molecular orbitals cannot be deduced from symmetry considerations.

The number of bonding, or antibonding, orbitals developed is equal to the number of atomic orbitals that are used. This follows from the fact that each atomic orbital is a description of an allowed electron behavior. As the atoms join to form a molecule, the details of the descriptions change but the number of these allowed behaviors does not. Thus, if one orbital of a given symmetry type is produced by the central atom and one of this type by the attached atoms, bonding and antibonding orbitals will be formed. The interaction splits the orbitals into high and low-energy orbitals; it does not change the number of orbitals.

Now let us have a look at CH_4 molecule; the figure indicates clearly the symmetries of the molecular orbitals of the CH_4 molecule.

METALLIC BONDING

The peculiar type of bonding which holds the atoms together in metal crystal is called the metallic Bonding.

To understand metallic bonding let us go through, '*Electron Sea Model*'.

Low ionization energies which implies that the valence electrons in metal atoms can easily be separated.

A number of vacant electron orbitals in their outermost shell. For example, the magnesium atom with the electron configuration $1s^2 2s^2 2p^6 3s^2 3p^0$, has three vacant 3p orbitals in its outer electron shell.

There is considerable overlapping of vacant orbitals on one atom with similar orbitals of adjacent atoms, throughout the metal crystal. Thus, it is possible for an electron to be delocalized and move freely in the vacant molecular orbital encompassing the entire metal crystal. The delocalized electrons no longer belong to individual metal atoms but rather to the crystal as a whole.

As a result of the delocalization of valence electrons, the positive metal ions that are produced, remain fixed in the crystal lattice while the delocalized electrons are free to move about in the vacant space in between. The metal is thus pictured as a network or lattice of positive ions of the metal immersed in a 'sea of electrons' or 'gas of electrons'. This relatively simple model of metallic bonding s referred to as the *Electron Sea model* or the *Electron Gas model*.

The electron sea model of metallic bonding explains fairly well the most characteristic physical properties of metals.

Luster or Reflectivity : The delocalized mobile electrons of the 'electron sea' account for this property. Light energy is absorbed by these electrons which jump into higher energy levels and return immediately to the ground level. In doing so, the electrons emit electromagnetic radiation (light) of the same frequency. Since the radiated energy is of same frequency as the incident light, we see it as a reflection of the original light.

Electric Conductivity : Another characteristic of metals is that they are good conductors of electricity. According to the electron sea model, the mobile electrons are free to move through the vacant space between metal ions. When electric voltage is applied at the two ends of a metal wire, it causes the electrons to be displaced in a given direction. The best conductors are the metals which attract their outer electrons the least (low ionization energy) and thus allow them the greatest freedom of movement.

Ductility and Malleability : The ductility and malleability of metals can also be explained by the electron sea model. In metals the positive ions are surrounded by the sea of electrons that 'flows' around them. If one layer of metal ions is forced across another, say by hammering, the internal structure remains essentially unchanged.

Electron Emission : When enough heat energy is applied to a metal to overcome the attraction between the positive metal ions and an outer electron, the electron is emitted from the metallic atom. When the frequency and, therefore, the energy of the light that strikes the metal is great enough to overcome the attractive forces, the electron escapes from the metal with a resultant decrease in the energy of the incident photon (Photoelectric effect).

The valence bonds that hold the atoms in a metal crystal together are not ionic, nor are they simply covalent in nature. Ionic bonding is obviously impossible here since all the atoms would tend to give electrons but none are willing to accept them. Ordinary covalent bonding is also ruled out as, for example, sodium atom with only one outer-shell electron could not be expected to form covalent bonds with 8 nearest neighbouring atoms in its crystal.

The Electron Sea Model is the most simplest one which explains precisely about metallic bonding.

ELECTRON NEGATIVITY

The bonding in many heteronuclear diatomic molecules is intermediate between the equally shared and ionic extremes.

In view of this unequal sharing of bond electrons, it seems desirable to try to assign to each atom a number which measures the tendency of the bonding electrons to be drawn toward that atom. *Electronegativity* is the name given to the index that attempts to represent this electron-attracting tendency. Several methods of arriving at electronegativity values have been suggested, but only two need be mentioned.

A rather direct, but somewhat limited, method due to R. S. Mulliken makes use of the ionization energy and the electron-affinity data. The attraction that an atom, or really the ion, exerts on a pair of electrons in a bond between that atom and another atom can be expected to be some average of the attraction of the free ion for an electron, *i.e.*, the ionization potential, and the attraction of the neutral atom for an electron, *i.e.*, the electron affinity. Numerical values are obtained that coincide with values from other methods if electronegativities, designated x, are calculated from x = (I + A)/540, where *I* and A are the ionization energy and electron affinity in kilojoules. In this way, for example, one calculates the electronegativity of chlorine as

$$x_{Cl} = \frac{125 + 325}{540} = 3.0$$

Similarly, values can be obtained for other elements for which ionization-potential and electron-affinity data are available.

An alternative method, due to Pauling, makes use of bond energies and a treatment of bond energies that is influenced by the variation theorem.

A description of the bonding in a heteronuclear bond, which allows for the ionic character of the bond, is expected, in view of the variation theorem, to lead to a better, *i.e.*, greater, calculated bond energy than would an incomplete, pure covalent-bond description.

An important step was made by Pauling in suggesting that the extra energy corresponding to the better description of the bonding electrons could be deduced from bond energies. In this method, the energy a bond would have if the electrons were equally distributed among the nuclei is calculated from the average of the covalent-bond energies of the atoms of the bond.

Thus, this hypothetical covalent-bond energy is calculated for HCl, as the average of the H_2 and Cl_2 bond energies. The actual bond energy is identified with the more complete description. Comparison of the values that are established, 428 and 336 kJ, confirms our ideas that a pure covalent description for the bond of a heteronuclear molecule is inadequate and that a good description of a heteronuclear chemical bond must include an ionic term. For almost all heteronuclear bonds a similar result is found, namely, that the actual energy is greater than the calculated covalent value. The calculated energy difference is represented by A.

Pauling found that a self-consistent set of electronegativities could best be deduced if one used the relation

$$x_B - x_A = \sqrt{\Delta}$$

with A expressed in electronvolts, where 1 eV = 96.49 kJ mol^{-1}. The use of the square root is quite arbitrary but leads, for example, to essentially the same value for $x_{Cl} - x_I$ from the data for ICl as from the data for HCl and HI.

Pauling's method allows differences in electronegativities to be deduced from bond-energy data. To obtain values for the individual atoms, it is necessary to pick an arbitrary reference point. If x_H is taken as 2.2, electronegativity values range from about 1.0 to 4.0, and this is considered a convenient range. Results obtained in this way are in general agreement with the results from Mulliken's method.

VIRIAL THEOREM

The kinetic and potential-energy components of chemical bonds are related by the virial theorem.

One basis for unscrambling the contributions to the total energy is the virial theorem. The theorem says that for any isolated, stationary-state collection of particles for which the forces follow an inverse-square force law, as electrostatic forces do, the relation

$$2(KE) = -U$$

holds; *i.e.,* twice the kinetic energy is equal but of opposite sign to the potential energy.

The total energy of the collection of particles, which, for example, could be the two protons and the one or two electrons of the H or H_2 species, is

$$\varepsilon = KE + U$$

which, with $2(KE) = -U$, becomes

$$\varepsilon = 1/2\ U$$

or

$$\varepsilon = -KE$$

These remarkable and simple relations can be applied immediately to the equilibrium condition of an atom or of a molecule. For H_2, for example, the net value of s is –458 kJ mol^{-1}. Thus, at this equilibrium position $(KE)H_2$ = +458 kJ mol^{-1} and U_{H2} = –916 kJ mol^{-1}, both relative to two hydrogen atoms. We see that the bond energy in H_2 results from the very large decrease in potential energy. This effect more than compensates for the increase in kinetic energy that results from the confinement of the electrons to these relatively small regions. Similar insights can be gained by applying the virial theorem to other systems in their equilibrium configuration.

Now let us see how we can interpret the energy of a chemical bond at internuclear separations other than that for the minimum in the binding-energy curve. Again H_2 will serve as an example. Consider a coordinate system centered on one atom of an H_2 molecule. The second atom will be at some distance r, the internuclear separation. For this distance to be other than the equilibrium one, some external force must act, and in acting, it contributes an additional potential-energy term to the molecule. The force that must be exerted is just the derivative of the total energy with respect to internuclear distance; *i.e.,* it is the slope of the experimentally based curve. The potential energy corresponding to this force is just r dε/dr, and thus the virial theorem becomes

$$2(KE) = -U - r\frac{d\varepsilon}{dr}$$

The kinetic and potential-energy contributions to the total energy e are now deduced from ε = KE + U.

$$KE = \varepsilon - r\frac{d\varepsilon}{dr}$$

and

$$U = 2\varepsilon + r\frac{d\varepsilon}{dr}$$

Since experimental e-versus-r curves like can be deduced, the derivative terms can be evaluated. Thus K.E-versus-r and U-versus-r curves can be constructed. Now you see that as the bond begins to form at relatively large internuclear distances, it is the kinetic energy that contributes to the binding, a result of the somewhat greater region in which the electrons can move. At this stage, additional electron-electron and proton-proton repulsions work to oppose the bonding. Only at shorter internuclear distances does the reverse occur.

THE PYRAMIDAL AND ANGULAR MOLECULES

Pyramidal Molecules

Ammonia Molecule : The Lewis structure of NH_3 shows that the central N atom has three bonding electrons and one lone electron pair. The VSEPR theory says that these electron pairs are directed to the corners of a tetrahedron. Thus we predict that H—N—H bond angle should be 109.5°. But the shape of a molecule is determined by the arrangement of atoms and not the unshared electrons. Thus, if we see only at the atoms, we can visualise NH_3 molecule as a pyramid with the N atom located at the apex and H atoms at the three corners of the triangular base.

According to VSEPR theory, a lone pair exerts greater repulsion on the bonding electron pairs than the bonding pairs do on each other. As a result, the bonds of NH_3 molecule are pushed slightly closer. This explains why the observed bond angle H—N—H is found to be 107.3° instead of 109.5° predicted from tetrahedral geometry.

Water Molecule

(a) In the structural formula of H_2O, the O atom is bonded to two H atoms by covalent bonds and has two lone pairs. Thus, O is surrounded by two bonding electron pairs and two unshared electron pairs. VSEPR theory says that in order to secure maximum separation between them, the four electron pairs are directed to the corners of a tetrahedron. If we look at the three atoms (and ignore the unshared pairs), the atoms HOH lie in the same plane and the predicted bond angle is 109.5°. But with two unshared pairs repelling the bonding pairs, the bond angle is compressed to 105°, the experimental value. Thus, the H_2O molecule is flat and bent at an angle at the O atom. Such a molecule is called a bent molecule or angular molecule.

(b) *Sulphur dioxide, SO_2* : The Lewis structure of SO_2 is given below. The S atom is bonded to one O by a double bond and to the other 0 by a single bond. It has an unshared electron pair. In VSEPR model a double bond is counted as a single electron pair. That way, the S atom is surrounded by three electron pairs, two bonding pairs and one unshared pair. For maximum separation the three electron pairs are directed to the corners of an equilateral triangle. The predicted bond angle is 120°. But with the unshared electron pair repelling the bonding electron pairs, the bond angle is actually reduced somewhat. Thus, SO_2 has a planar bent molecule with the observed bond angle 119.5°.

MECHANISMS OF CHEMICAL REACTIONS

In the equilibrium reactions, the rates of the two opposing reactions are equal and the concentrations of reactants or products do not change with lapse of time. But most chemical reactions are *spontaneous reactions*. These reactions occur from left to right till all the reactants are converted to products. A spontaneous reaction may be slow or it may be fast. For example, the reaction between aqueous sodium chloride and silver nitrate is a fast reaction. The precipitate of AgCl is formed as fast as $AgNO_3$ solution is added to NaCl solution. On the contrary, the rusting of iron is a slow reaction that occurs over the years.

The branch of Physical chemistry which deals with the rate of reactions is called Chemical Kinetics.

The study of Chemical Kinetics includes :

(1) The rate of a reaction and rate laws.

(2) The factors as temperature, pressure, concentration and catalyst, that influence the rate of a reaction.

(3) The mechanism or the sequence of steps by which a reaction occurs.

The knowledge of the rate of reactions is very valuable to understand the chemistry of reactions. It is also of great importance in selecting optimum conditions for an industrial process so that it proceeds at a rate to give maximum yield.

Chemical systems at equilibrium have been treated, and the nature of chemical substances, which may be reactants or products in a chemical reaction, has been studied.

Now the actual process of chemical reactions is investigated, and our attention is focused not only on the reactants and the products but also on the details of the transformation from one set of chemical species to another. That this aspect has previously been neglected or avoided is emphasized by recognizing that the time variable has so far been absent and that it will now play a major role.

The question of how reactants are converted to products in some particular chemical reaction calls for an answer in terms of molecular-level happenings. Two phases of the question and the answer can be recognized. One part focuses on the molecular details of what are called *elementary reactions*, *e.g.*, the collision of two molecules to produce one or more new species.

The other part of kinetics studies deals with the sequence of elementary reactions that constitutes the overall chemical transformation. Information comes from a variety of experimental approaches but principally from measurements of how the rate of the reaction depends on the concentrations of the reagents. Such experimental results are usually expressed analytically by the *rate equation*, which describes the dependence of the rate of reaction at a given temperature on the concentration of the reagents.

Most reactions appear to be the result of a succession of molecular events *e.g.*, collisions or rearrangements. Before we tackle the details of the elementary process, let us see how we deduce the sequences of the steps, the *mechanism* for a particular reaction.

The unraveling of these steps is based primarily on the dependence of the rate of the reaction on the concentration of various reactants and products. Therefore, we begin by turning our attention to the study of the rates of chemical reactions and the concentration dependence of these rates.

THE RATE OF REACTION

The dependence of the rate of a reaction on the concentrations of the reagents is expressed by a rate equation.

Studies of the rate of a reaction usually give information on the decrease in the amount of one of the reactants or on the increase in the amount of a product that occurs in some time interval. If the reaction system is one of constant or near constant volume, the change in the amount of reagent will correspond to a change in the concentration of that reagent. For liquid systems the rate of a reaction is often expressed in terms of the rate of change of the molar concentration of a reagent.

The rate of a reaction tells as to what speed the reaction occurs. Let us consider a simple reaction

$$A \rightarrow B$$

The concentration of the reactant A decreases and that of B increases as time passes. The rate of *reaction* is defined as *the change in concentration of any of reactant or products per unit time.* For the given reaction the rate of reaction may be equal to the rate of disappearance of A which is equal to the rate of appearance of B.

Thus

$$\begin{aligned}\text{rate of reaction} &= \text{rate of disappearance of A} \\ &= \text{rate of appearance of B.}\end{aligned}$$

or

$$\text{rate} = -\frac{d[A]}{dt}$$

$$+\frac{d[B]}{dt}$$

where [ ] represents the concentration in moles per litre whereas d represents infinitesimally small change in concentration. Negative sign shows the concentration of the reactant A decreases whereas the positive sign indicates the increase in concentration of the product B.

The rate of reaction, generally depends on the concentrations of one or more of the reactants. Thus the rate changes as the reaction proceeds. This complication can be avoided by studying "initial rates." These are the rates of the reaction during the initial stages when the concentrations of the reactants have not changed appreciably. Reaction rate has the units of concentration divided by time. We express concentrations in moles per litre (mol/litre or mol/l ormol l^{-1}) but time may be given in any convenient unit second (s), minutes (min), hours (h), days (d) or possibly years. Therefore, the units of reaction rates may be

mole/litre sec or mol l^{-1} s^{-1}

mole/litre min or mol l^{-1} min^{-1}

ORDER OF A REACTION

The order of a reaction is defined as the slim of the powers of concentrations in the rate law

Let us consider the example of a reaction which has the rate law

$$rate = k[A]^m [B]^n$$

The order of such a reaction is (m + n).

The order of a reaction can also be defined with respect to a single reactant Thus the reaction order with respect to A is m and with respect to B it is n. The *overall order of reaction* (m + n) may range from 1 to 3 and can be fractional.

Reactions may be classified according to the order.

m + n = 1, it is first order reaction

m + n = 2, it is second order reaction

m + n = 3, it is third order reaction.

Zero Order Reaction

A reactant whose concentration does not affect the reaction rate is not included in the rate law. In effect, the concentration of such a reactant has the power 0. Thus $[A]^0 = 1$. A zero order reaction is one whose rate is independent of concentration.

For example, the rate law for the reaction

$$NO_2 + CO \rightarrow NO + CO_2$$

at 200°C is ate $= k[NO_2]^2$

Here the rate does not depend on [CO], so this is not included in the rate law and the power of [CO] is understood to be zero. The reaction is zeroth order with respect to CO. The reaction is second order with respect to $[NO_2]$. The overall reaction order is 2 + 0 = 2.

MOLECULARITY OF A REACTION

Chemical reactions may be classed into two types;

(a) Elementary reactions

(b) Complex reactions

An *elementary reaction* is a simple reaction which occurs in a single step.

A complex reaction is that which occurs in two or more steps.

Molecularity of an Elementary reaction

The molecularity of an elementary reaction is defined as; the number of reactant molecules involved in a reaction. Thus, the molecularity of an elementary reaction is 1, 2, 3, etc., according as one, two or three reactant molecules are participating in the reaction. The elementary reactions having molecularity 1, 2 and 3 are called *unimolecular, bimolecular and trimolecular* respectively.

Differences between order and molecularity

Order of a Reaction	Molecularity of a Reaction
1. It is the *sum of powers* of the concentration terms in the rate law expression.	1. It is the number of *reacting spices* undergoing simultaneous collision in the elementary or simple reaction.
2. It is an *experimentally* determined value.	2. It is a *theoretical* concept.
3. It can have *fractional* value.	3. It is always a *whole number*.
4. It can assume *zero value*.	4. It *can not have zero value*.
5. Order of a reaction *can change* with the conditions such as pressure, temperature, concentration.	5. Molecularity is *invariant* for a chemical equation.

FIRST ORDER REACTIONS

A first-order reaction is one for which, at a given temperature, the rate of the reaction depends only on the first power of the concentration of a single reacting species. If the concentration of this species is represented by c (for solutions, the units of moles per liter are ordinarily used), and if the volume of the system remains essentially constant during the course of the reaction, the first-order rate equation can be written.

Let us consider a first order reaction

$$A \rightarrow \text{products.}$$

Suppose that at the beginning of the reaction (t = 0), the concentration of A is a moles litre $^{-1}$. If after time t, x moles of A have changed, the concentration of A is a – x. We know that for a first order reaction, the rate of reaction, dx/dt, is directly proportional to the concentration of the reactant. Thus,

$$\frac{dx}{dt} = k(a-x)$$

or

$$\frac{dx}{a-x} = kdt$$

Integration of the expression (1) gives

$$\int \frac{dx}{a-x} = \int kdt$$

or

$$-\ln(a-x) = kt + I$$

where I is the constant of integration. The constant k may be evaluated by putting t = 0 and x = 0.

Thus, $$I = -\ln a$$

Substituting for I in equation (2)

$$\ln \frac{a}{a-x} = kt$$

or

$$k = \frac{1}{t} \ln \frac{a}{a-x}$$

Changing into common logarithms

$$k = \frac{2.303}{t} \log \frac{a}{a-x}$$

The value of k can be found by substituting the values of a and (a – x) determined experimentally at time interval t during the course of the reaction.

Sometimes the integrated rate law in the following form is also used:

$$k = \frac{2.303}{t_2 - t_1} \log \frac{(a - x_1)}{(a - x_2)}$$

where x_1 and x are the amounts decomposed at time intervals t_1 and t_2 respectively from the start.

Decomposition of Nitrogen pentoxide in carbon tetrachloride solution is a good example for, first order reaction.

Now let us see some of the examples of first order reaction.

Decomposition of H_2O_2 in Aqueous Solution : The decomposition of H_2O_2 in the presence of Pt as catalyst is a first order reaction.

$$H_2O_2 \xrightarrow{Pt} H_2O + O$$

The progress of the reaction is followed by titrating equal volumes of the reaction mixture against standard $KMnO_4$ solution at different time intervals.

Example : A solution of H_2O_2 when titrated against $KMnO_4$ solution at different time intervals gave the following results :

r (minutes)	0	10	20
Vol $KMnO_4$ used for 10 ml H_2SO_2	23.8ml	14.7ml	9.1ml

Hydrolysis of an Ester : The hydrolysis of ethyl acetate or methyl acetate in the presence of mineral acid as catalyst, is a first order reaction.

$$\underset{\text{ethyl acetate}}{CH_3COOC_2H_5} + H_2O \xrightarrow{H^+} \underset{\text{acetic acid}}{CH_3COOH} + C_2H_5OH$$

For studying the kinetics of the reaction, a known volume of ethyl acetate is mixed with a relatively large quantity of acid solution, say N/2 HCl. At various intervals of time, a known volume of the reaction mixture is titrated against a standard alkali solution. Hydrolysis of the ester produces acetic acid. Therefore, as the reaction proceeds, the volume of alkali required for titation goes on increasing.

Inversion of Cane Sugar (Sucrose) : The inversion of cane sugar or sucrose catalysed with dil HCl.

$$C_{12}H_{22}O_{11} + H_2O \xrightarrow{H^+} \underset{\text{D-glucose}}{C_6H_{12}O_6} + \underset{\text{D-fructose}}{C_6H_{12}O_6}$$

follows the first order kinetics. The progress of the reaction is followed by noting the optical rotation of the reaction mixture with the help of a polarimeter at different time intervals. The optical rotation goes on changing since D-glucose rotates the plane of polarised light to the right and D-fructose to the left. *The change in rotation is proportional to the amount of sugar decomposed.*

We can show that inversion of sucrose is a first order reaction.

The available data is substituted in the first order rate equation for different time intervals.

$$k = \frac{2.303}{t} \log \frac{r_0 - r_\infty}{t_t - r_\infty}$$

$r_o - r_\infty = 24.09 - (-10.74) = 34.83$ for all time intervals. Thus, the value of rate constant can be found.

$$\text{time (t)} \qquad r_t = r_\infty \; k = \frac{1}{t} \log \frac{(r_0 - r_\infty)}{(t_t - r_\infty)}$$

$$32.14 \; k = \frac{1}{7.18} \log \frac{34.83}{32.14} = 0.0047$$

$$28.44 \; k = \frac{1}{18} \log \frac{34.83}{28.44} = 0.0048$$

$$25.74 \; k = \frac{1}{27.1} \log \frac{34.83}{25.74} = 0.0048$$

Since the value of k comes out to be constant, the inversion of sucrose is a first order reaction.

SECOND ORDER REACTION

A reaction is classified as second-order if the rate of the reaction is proportional to the square of the concentration of one of the reagents or to the product of the concentrations of two species of the reagents. The second situation leads to the same equations as the first if the two reactants are used up at the same rate and if their initial concentrations are equal.

Rate = kc^2 : For these situations, the rate equation is

$$-\frac{dc}{dt} = kc^2$$

where c is the concentration of the single reagent or of one of the two reagents. Again the kinetic data are usually compared with the integrated form of the equation. One has

$$-\int_{co}^{c} \frac{dc}{c^2} = k\int_0^t dt$$

and
$$\frac{1}{c} - \frac{1}{c_0} = kt$$

A reaction of the types considered so far is therefore second-order if a plot of 1/c versus t gives a straight line. The slope of the straight line is equal to the rate constant. This constant involves the units of concentration, differing in this respect from the first-order rate constant that involves only the units of time. Furthermore, the time for the concentration to drop to half its initial value is deduced like this,

$$t_{1/2} = \frac{1}{kc_0}$$

The half-life therefore depends on the initial concentration and is not a convenient way of expressing the rate constant of second-order reactions.

Let us consider

A + B → products, and the initial concentrations of A and B are equal, then the kinetic data, can be treated in terms of the following quantities,

a = initial conc. of A

b = initial conc. of B.

x = decrease in A or B at time.

t = amount of product at time t.

a–x = concentration of A at time t.

b–x = concentration of B at time t.

Then the differential second order rate equation will be,

$$\frac{dx}{dt} = k[A][B] = k\,(a-x)(b-x).$$

The integration can be performed by using partial fractions.

so, $$\frac{dx}{(a-x)(b-x)} = kdt$$

$$\frac{1}{a-b}\int_0^x\left(-\frac{dx}{a-x}+\frac{dx}{b-x}\right) = k\int_0^t dt$$

On integration this gives

$$\frac{1}{a-b}\left[\text{In }(a-x) - \text{In }(b-x)\right]_0^x = kt$$

Insertion of the limits and rearrangement give, finally,

$$\frac{1}{a-b}\ \text{In}\ \frac{b(a-x)}{a(b-x)} = kt$$

THIRD ORDER REACTIONS

Let us consider a simple third order reaction of the type

$$3A \rightarrow \text{products.}$$

Let the initial concentration of A be a moles litre^{-1} and after time t, x moles have reacted. Therefore, the concentration of A becomes (a – x). The rate law may be written as:

$$\frac{dx}{dt} = k(a-x)^3 \qquad ...(1)$$

Rearranging equation (1), we have

$$\frac{dx}{(a-x)^3} = kdt \qquad ...(2)$$

On integration, it gives

$$\frac{1}{2(a-x)^2} = kt + I \qquad ...(3)$$

wh]ere I is the integration constant. I can be evaluated by putting x = 0 and t = 0 Thus,

$$I = \frac{1}{2a^2}$$

By substituting the value of I in (3), we can write

$$kt = \frac{1}{2(a-x)^2} - \frac{1}{2a^2}$$

Therefore, $$k = \frac{1}{t} \cdot \frac{x(2a - x)}{2a^2 (a - x)^2}$$

This is the integrated rate equation for a third order reaction.

There are not many reactions showing third order kinetics. A few of the known examples are:

(i) $2FeCl_3(aq) + SnCl_2aq) \rightarrow 2Fe_2Cl_2 + SnCl_4$

(ii) $2NO_{(g)} + O_{2(g)} \rightarrow 2NO_{2(g)}$

(iii) $2NO_{(g)} + Cl_{2(g)} \rightarrow 2NOCl_{(g)}$

Units of Third Order Rate Constant

The rate constant for a third order reaction is

$$k = \frac{1}{t} \cdot \frac{x(2a - x)}{2a^2 (a - x)^2}$$

or $$k = \frac{\text{concentration} \times \text{concentration}}{(\text{concentration})^2 \ (\text{concentration})^2} \times \frac{1}{\text{time}}$$

$$= \frac{1}{(\text{concentration})^2} \times \frac{1}{\text{time}}$$

$$= \frac{1}{(\text{mol/litre})^2} \times \frac{1}{\text{time}}$$

Thus the units of k for a third order reaction are

$mol^{-2} l^2 time^{-1}$.

Half-Life of a Reaction

Reaction rates cab also be expressed in terms of *half-life* or *half-life period*. It is defined as: *the time required for the concentration of a reactant to decrease to half its initial value.*

In other words, half-life is the time required for one-half of the reaction to be completed. It is represented by the symbol $t_{1/2}$ or $t_{0.5}$.

The half-life of a reaction can be calculated by using the integrated rate equation for its order.

Calculation of Half-Life of a First Order Reaction

The integrated rate equation (4) for a first order reaction can be stated as:

$$k = \frac{2.303}{t} \log \frac{[A]_0}{[A]}$$

where $[A]_0$ is initial concentration and [A] is concentration at any time t. Half-life, $t_{1/2}$, is time when initial concentration reduces to 1/2 *i.e.,*

$$[A] = 1/2\ [A]_0$$

Substituting values in the integrated rate equation, we have

$$k = \frac{2.303}{t_{1/2}} \log \frac{[A]_0}{\frac{1}{2}[A]_0} = \frac{2.303}{t_{1/2}} \log 2$$

or $$t_{1/2} = \frac{2.303}{k} \log 2 = \frac{2.303}{k} \times 0.3010$$

or $$t_{1/2} = \frac{0.693}{k}$$

It is clear from this relation that:

(1) half-life for a first order reaction is independent of the initial concentration.

(2) it is inversely proportional to k, the rate-constant.

As in case of a first order reaction, half-life for a second order reaction is inversely proportional to rate constant k. While half-life of a first order reaction is independent of initial concentration, *half-life of a second order reaction depends on initial concentration.* This fact can be used to distinguish between a first order and a second order reaction.

DETERMINATION OF THE ORDER OF A REACTION

We can use many types (at least four) methods to determine the order of a reaction. Let us study some of the reactions.

Using Integrated Rate Equations

The reaction under study is performed by taking different initial concentrations of the reactant (a) and noting the concentration (a – x) after regular time intervals (t). The experimental values of a, (a – x) and t are then substituted into the integrated rate equations for the first, second and third order reactions. The rate equation which yields a constant value of A; corresponds to the correct order of the reaction. This method of ascertaining the order of a reaction is essentially a method of hit-and-trial but was the first to be employed. It is still used extensively to find the order of simple reactions.

Graphical Method

For reactions of the type A → products, we can determine the reaction order by seeing whether a graph of the data fits one of the integrated rate equations

In Case of First Order

We have already derived the integrated rate equation for first order as

$$\ln \frac{a}{a - x} = kt$$

Simplifying, it becomes

$$\begin{array}{lll} \ln (a - x) & = -kt & + \ln a \\ \uparrow & \quad\uparrow & \quad\uparrow \\ y & = mx & + \; b \end{array}$$

Thus, the two variables in the first order rate equation are:

$$\ln \frac{a}{a - x} = \text{and } t$$

Hence, if $\ln \frac{a}{a - x}$ is plotted against t and straight line results the corresponding reaction is of the first order.

In Case of Second Order

Second order rate equation can be written as

$$\begin{array}{lll} \frac{1}{a - x} & = kt & + \frac{1}{a} \\ \uparrow & \quad\uparrow & \quad\uparrow \\ y & = mx & + \; b \end{array}$$

This is the equation of a straight line, y = mx + b. Here the two variables are

$$\frac{1}{a - x} \text{ and } t$$

Thus, when $\frac{1}{a - x}$ is plotted against t and we get a straight line the reaction is second order. In case a curve is obtained, the reaction is not second order.

Using Half-life Period

Two separate experiments are performed by taking different initial concentrations of reactants. The progress of the reaction in each case is recorded by analysis. When the initial concentration is reduced to one-half, the time is noted. Let the initial concentrations in the two experiments be $[A_1]$ and $[A_2]$, while times for completion of half change are t_1 and t_2 respectively.

Ostwald's Isolation Method

This method is employed in determining the order of complicated reactions by 'isolating' one of the reactants so far as its influence on (he rate of reaction is concerned. Suppose the reaction under consideration is:

$$A + B + \rightarrow \text{products.}$$

The order of the reaction with respect to A, B and C is determined. For the determination of the order of reaction with respect to A, B and C are taken in a large excess so that their concentrations are not affected during the reaction.

The order of the reaction is then determined by using any of the methods described earlier. Likewise, the order of the reaction with respect to B and C is determined. In n_A, n_B and n_C are the orders of the reaction with respect to A, B and C respectively, the order of the reaction n is given by the expression

$$n = n_A + n_B + n_C$$

COMPLEX REACTIONS

The reactions which proceed in a series of steps instead of a single step are known as complex reactions. These reactions are generally known as simultaneous reactions. Now let us study the types of complex reactions.

Consecutive Reactions

Parallel or Side Reactions

Reversible or Opposing Reactions

Consecutive Reactions

The reactions in which the final product is formed through one or more intermediate steps are called *consecutive reactions*. These are also known as *sequential* reactions. In such reactions the product formed in one of the elementary reactions acts as the reactant for some other elementary reaction. Various step reactions can be written for the overall reaction as shown below :

$$A \xrightarrow{k_1} B \xrightarrow{k_2} C$$

	A	B	C
Initial Cone.	[A]g	o	o
Cone. after time t	[A]	[B]	[C]

In the above reaction the product C is formed from the reactant A through intermediate B. In this reaction each stage has its own different rate constants, k_1, for the first step and k_2 for the second step. The net or overall rate of reaction depends upon the magnitude of these two rate constants. The initial concentration and concentration after time t are shown below each species in above reaction under consideration.

It is clear that

$$[A]_0 = [A] + [B] + [C]$$

The differential rate expressions are

$$\frac{-d[A]}{dT} = k_1[A]$$

$$\frac{d[B]}{dT} = k_1[A] - k_2[B]$$

and $$\frac{d[C]}{dT} = k_2[B]$$

The concentration of A decreases exponentially, the concentration of B first increases and then decreases and that of C increases (from zero) with time and finally attains the value equal to $[A]_0$ (initial concentration A) when all A has changed into the final product C.

Decomposition of Ethylene oxide and decomposition of dimethyl ether in gaseous phase are the good examples for first order consecutive reactions.

Parallel or Side Reactions

In these reactions the reacting substance follows two or more paths to give two or more products. The preferential rate of such reaction may be changed by varying the conditions like pressures temperature or catalyst. The reaction in which the maximum yield of the product is obtained is called the main or major reaction while the other reaction (or reactions) are called side or parallel reactions. In the above reaction the reactant A gives two products B and C separately in two different reactions with rate constants k_1 and k_2 respectively. If $k_1 > k_2$ the reaction $A \rightarrow B$ will be the major reaction and $A \rightarrow C$ will be the side or parallel reaction. Let us assume that both these reactions are of first order and concentration of A is [A] at the time t. The differential rate expressions are

$$r_1 = \frac{-d[A]}{dT} = k_1[A] \quad ...(i)$$

and

$$r_2 = \frac{-d[A]}{dT} = k_2[A] \quad ...(ii)$$

The total rate of disappearance of A is given by

$$\frac{-d[A]}{dT} = r_1 + r_2 = k_1[A] + k_2[A]$$

$$= (k_1 + k_2)[A]$$

$$= k'[A],$$

where k' is the first order rate constant. It is equal to the sum of the two rate constants k_1 and k_2 of two side reaction.

Integrating equation (iii), we get

$$\int \frac{-d[A]}{dT} = \int k'[A]$$

Applying the limits $[A_0]$ and $[A]_t$ and O and t, we get,

$$\int_{[A]_0}^{[A]_t} \frac{-d[A]}{dt} = l' \int_0^t dt$$

$$\text{In} \frac{[A]_0}{[A]_t} = k't + (k_1 + k_2)_t$$

by equations (i) and (ii) we get,

$$\frac{r_1}{r_2} = \frac{k_1[A]}{k_2[A]} = \frac{k_1}{k_2}$$

REVERSIBLE OR OPPOSING REACTIONS

In reversible or opposing reactions the products formed also react to give back the reactants. Initially, the rate of forward reaction is very large which decreases with passage of time and the rate of backward or reverse reaction is zero which increases with passage of time. A stage is reached when two rates become equal. This situation is called the chemical equilibrium. It is dynamic in nature *i.e.*, all the species are reaching at the rate at which they are being formed. A reaction of this type may be represented as

$$A \underset{k_b}{\overset{k_f}{\rightleftharpoons}} B$$

Initial concentration	$[A]_0$	0
Cone. after time t	[A]	[B]

Where k_f and k_b are the rate constants of the forward and backward reactions respectively.

The overall rate of reaction is given by

Rate of Reaction = Rate of forward reaction – Rate of backward reaction

i.e., $$\frac{-d[A]}{dt} = \frac{d[B]}{dt} = k_f[A] - k_b[B] \quad ...(i)$$

If $[A]_0$ is the initial concentration of A and x moles of it have reacted in time t

then $[A]_t = [A]o - x$

and $[B] = x$

Substituting these is equation (i), we get

$$\frac{dx}{dt} = k_f\,([A]_0 - x) - k_b x \quad ...(ii)$$

At equilibrium $\frac{dx}{dt} = 0$

Hence $k_f\,([A]_0 - x_{eq}) = k_b\,x_{eq}$ ...(iii)

where x_{eq} is the concentration of A that has reacted into B at equilibrium. From equation (iii) we have

$$k_b = k_f\left(\frac{[A]_0 - x_{eq}}{x_{eq}}\right)$$

Substituting the value of k_b in equation (ii), we get

$$\frac{dx}{dt} = k_f([A]_0 - x) - k_f\left(\frac{[A]_0 - x_{eq}}{x_{eq}}\right)x$$

Integrating this equation between the limits

$$t = 0, \; x = 0$$

and $$t = t \; x = x_{eq}$$

We have

$$\int_0^x \frac{dx}{x_{eq} - x} = k_f \frac{[A_0]}{x_{eq}} \int_0^t dt$$

$$= -\ln(x_{eq} - x) + \ln x_{eq}. = k_f \frac{[A]_0 t}{x_{eq}}$$

or $$\ln \frac{x_{eq}}{x_{eq} - x} = k_f \frac{[A]_0}{x_{eq}} t$$

From this equation we can find the value of k_f from the quantities $[A]_0$, x_{eq}, and x at time t. All these quantities can be measured easily. From the value of k_f the value of k_b can be calculated by using the

$$k_b = k_f \left[\frac{[A_0] - x_{eq}}{x_{eq}} \right].$$

CATALYSIS

Catalyst is defined as a substance which alters the rate of a chemical reaction, itself remaining chemically unchanged at the end of the reaction. The process is called *Catalysis.*

As evident from the above definition, a catalyst may increase or decrease the rate of a reaction.

A catalyst which enhances the rate of a reaction is called a *Positive catalyst* and the process *Positive catalysis* or simply *Catalysis.*

A catalyst which retards the rate of a reaction is called a *Negative catalyst* and the process *Negative catalysis.*

We will first proceed to discuss 'positive catalysis' or catalysis as it is commonly designated.

There are two main types of catalysis:

(a) Homogeneous catalysis

(b) Heterogeneous catalysis

Also, there is a third types of catalysis known as Enzyme catalysis which is largely of biological interest. This will be discussed separately at a later stage.

In homogeneous catalysis, the catalyst is in the same phase as the reactants and is evenly distributed throughout. This type of catalysis can occur in gas phase or the liquid (solution) phase.

(i) Oxidation of sulphur dioxide (SO_2) to sulphur trioxide (SO_3) with nitric oxide (NO) as catalyst,

$$\underset{\text{gas}}{2SO_2} + \underset{\text{gas}}{O_2} + \underset{\text{gas}}{[NO]} \rightarrow \underset{\text{gas}}{2SO_3} + [NO]$$

(ii) Decomposition of acetaldehyde (CH_3CHO) with iodine (I_2) catalyst,

$$\underset{\text{vapour}}{CH_3CHO} + \underset{\text{vapour}}{[I_2]} \rightarrow \underset{\text{gas}}{CH_4} + \underset{\text{gas}}{CO}$$

Catalysis in Solution Phase

Many reactions in solutions are catalysed by acids (H^+) and bases (OH^-).

(i) Hydrolysis of cane sugar in aqueous solution in the presence of mineral acid as catalyst,

$$\underset{\text{cane sugar}}{C_{12}H_{22}O_{11}} + H_2O \xrightarrow{H_2SO_4} \underset{\text{glucose}}{C_6H_{12}O_6} + \underset{\text{fructose}}{C_6H_{12}O_6} + [H_2SO_4]$$

(ii) Hydrolysis of an ester in the presence of acid or alkali,

$$\underset{\text{ethyl acetate}}{CH_3COOC_2H_5} + H_2O \xrightarrow{H^+/OH} \underset{\text{acetic acid}}{CH_3COOH} + \underset{\text{ethanol}}{C_2H_5OH}$$

The catalysis in which the catalyst is in a different physical phase from the reactants is termed Heterogeneous catalysis.

Examples:

(in gaseous phase) $2SO_2 + O_2 + [Pt] \rightarrow 2SO_3 + [Pt]$

$N_2 + 3H_2 + [Fe] \rightarrow 2NH_3 + [Fe]$.

CATALYTIC REACTIONS

Now let us deal with certain characteristic features of catalytic reactions.

A catalyst remains unchanged in mass and chemical composition at the end of the reaction. Qualitative and quantitative analysis show that a catalyst undergoes no change in mass of chemical nature. However, it may undergo a physical change; Thus granular manganese dioxide (MnO_2) used as a catalyst in the thermal decomposition of potassium chlorate is left as a fine powder at the end of the reaction.

A small quantity of catalyst is generally needed to produce almost unlimited reaction.

Sometimes a trace of a metal catalyst is required to affect very large amounts of reactants. For example, one ten-millionth of its mass of finely divided platinum is all that is needed to catalyse the decomposition of hydrogen peroxide.

On the other hand, there are catalysts which, need to be present in relatively large amount to be effective. Thus, in Friedel-Crafts reaction.

$$C_6H_6 + C_2H_5Cl \xrightarrow{AlCl_3} C_6H_5C_2H_5 + HCl$$

anhydrous aluminium chloride functions as a catalyst effectively when present to the extent of 30 per cent of the mass of benzene.

For the acid and alkaline hydrolysis of an ester,

$$\underset{\text{ester}}{RCOOR'} + H_2O \xrightarrow{H^+ \text{ or } OH^-} RCOOH + R'OH$$

the rate of reaction is proportional to the concentration of the catalyst (H^+ or OH^-).

A catalyst is more effective when finely divided

In heterogeneous catalysis, the solid catalyst is more effective when in a state of fine subdivision than it is used in bulk. Thus, a lump of platinum will have much less catalytic activity than colloidal or platinised asbestos. Finely divided nickel is a better catalyst than lumps of solid nickel.

A catalyst is specific in its action

While a particular catalyst works for one reaction, it will not necessarily work for another reaction. Different catalysts, moreover, can bring about completely different reactions for the same substance. For example, ethanol (C_2H_5OH) gives ethene (C_2H_4) when passed over hot aluminium oxide,

$$C_2H_5OH \xrightarrow{Al_2O_3} \underset{\text{ethene}}{C_2H_4} + H_2O$$

but with hot copper it gives ethanal (CH_3CHO).

$$C_2H_5OH \xrightarrow{Cu} \underset{\text{ethanal}}{CH_3CHO} + H_2$$

A catalyst cannot, in general, initiate a reaction

In most cases a catalyst speeds up a reaction already in progress and does not initiate (or start) the r reaction. But there are certain reactions where the reactants do not combine for very long period (perhaps years). For example, a mixture of hydrogen and oxygen, which remains unchanged almost indefinitely at room temperature, can be brought to reaction by the catalyst *platinum black* in a few seconds.

$$H_2 + O_2 \xrightarrow{\text{room temp.}} \text{No reaction}$$

$$2H_2 + O_2 \xrightarrow{\text{Pt black}} 2H_2O$$

Thus, *it is now considered that the catalyst can initiate a reaction.* According to this view the reacting molecules (in the absence of catalyst) do not possess minimum kinetic energies tor successful collisions. The molecules rebound from collisions without reacting at all.

A catalyst does not affect the final position of equilibrium, although it shortens the time required to establish the equilibrium

It implies that in a reversible reaction the catalyst accelerates the forward and the reverse reactions equally. Thus the ratio of the rates of two opposing reactions *i.e.*, the equilibrium constant, remains unchanged.

The activity of a catalyst can often be increased by addition of a small quantity of a second material. This second substance is either not a catalyst itself for the reaction or it may be a feeble catalyst.

A substance which, though itself not a catalyst, promotes the activity of a catalyst is called a promoter.

Example of Promoters

Molybdenum (Mo) or aluminium oxide (Al_2O_3) promotes the activity of iron catalyst in the Haber synthesis for the manufacture of ammonia.

$$N_2 + 3H_2 \underset{+Mo}{\overset{Fe}{\rightleftharpoons}} 2NH_3$$

In some reactions, mixtures of catalysts are used to obtain the maximum catalytic efficiency. For example, in the synthesis of methanol

(CH_3OH) from carbon monoxide and hydrogen, a mixture of zinc and chromium oxide is used as a catalyst.

$$CO + 2H_2 \xrightarrow[Cr_2O_3]{ZnO} CH_3OH$$

The theory of promotion of a catalyst is not clearly understood. Presumably:

(1) *Change of Lattice Spacing* : The lattice spacing of the catalyst is changed thus enhancing the spaces between the catalyst particles. The adsorbed molecules of the reactant (say H_2) are further weakened and cleaved. This makes the reaction go faster.

(2) *Increase of Peaks and Cracks* : The presence of the promoter increases the peaks and cracks on the catalyst surface. This increases the concentration of the reactant molecules and hence the rate of reaction.

The phenomenon of promotion is a common feature of heterogeneous catalysis. A substance which destroys the activity of the catalyst to accelerate a reaction, is called a poison and the process is called Catalytic poisoning.

ENZYME CATALYSIS

Many enzyme-catalyzed reactions follow a complex rate equation that can be written in terms of the total amount of enzyme and the total amount of substrate in the reaction system.

The effect of enzymes on the rate with which chemical reactions move toward their equilibrium position provides one of the most dramatic catalytic effects. Much of the current interest in the subject is centered on the details of the interaction between the *enzyme*, which is the catalyst, and the material, known as the *substrate*, whose reaction it affects. It is important to understand how an enzyme-catalyzed reaction proceeds in time and how the catalytic activity of the enzyme-substrate pair is evaluated from the measurement of the progress of such reactions.

The experimental data for enzyme-catalyzed reactions show a variety of forms that depend on the enzyme, the substrate, the temperature, the presence of interfering substance and so forth.

The long chains of the enzyme (protein) molecules are coiled on each other to make a rigid colloidal particle with cavities on its surface. These cavities which are of characteristic shape and abound in active groups (NH_2, COOH, SH, OH), are termed *Active centres*. The molecules of substrate which have complementary shape, fit into these cavities just as key fits into a lock (*Lock-and-Key theory*). By virtue of the presence of active groups, the enzyme forms an activated complex with the substrate which at once decomposes to yield the products.

Enzyme are unique in their efficiency and high degree of specificity. They behave like inorganic heterogeneous catalysts. Now let us learn about some more features of enzyme catalysis.

Enzymes are the most efficient catalysts known

The enzyme catalysed reactions proceed at fantastic high rates in comparison to those catalysed by inorganic substances. Thus, one molecule of an enzyme may transform one million molecules of the substrate (reactant) per minute.

Like inorganic catalysts, enzymes function by lowering the activation energy of a reaction. For example, the activation energy of the decomposition of hydrogen peroxide,

$$2H_2O_2 \rightarrow 2H_2O + O_2$$

without a catalyst is 18 kcal/mole. With colloidal platinum (inorganic catalyst), the activation energy is lowered by 11.7 kcal/mole. The enzyme *catalse* lowers the activation energy of the same reaction to less than 2 kcal/mole.

Enzyme catalysis is marked by absolute specificity

An enzyme as a rule catalyses just one reaction with a particular substance. For example, urease (an enzyme derived from soya bean) catalyses the hydrolysis of urea and no other amide, not even methylurea.

The rate of enzyme catalysed reactions is maximum at the optimum temperature

The rate of an enzyme catalysed reaction is increased with the rise of temperature but up to a certain point. Thereafter the enzyme is denatured as its protein structure is gradually destroyed. Thus the rate of reaction drops and eventually becomes zero when the enzyme is completely destroyed. The rate of an enzyme reaction with raising of temperature gives a bell-shaped curve. The temperature at which the reaction rate is maximum is called the optimum temperature.

For example, the optimum temperature of enzyme reactions occurring in human body is 37°C (98.6°F). At much higher temperatures, all physiological reactions will cease due to loss of enzymatic activity. This is one reason why high body temperature (fever) is very dangerous.

Rate of enzyme catalysed reactions is maximum at the optimum pH

The rate of an enzyme catalysed reaction varies with pH of the system. The rate passes through a maximum at a particular pH, known as the optimum pH. The enzyme activity is lower at other values of pH. Thus many enzymes of the body function best at pH of about 7.4, the pH of the blood and body fluids.

THEORIES OF CATALYSIS

There are two main theories of catalysis:

(1) Intermediate Compound Formation theory

(2) The Adsorption theory

In general, the Intermediate Compound Formation theory applies to homogeneous catalytic P reactions and the Adsorption theory applies to heterogeneous catalytic reactions.

The Intermediate Compound Formation Theory

As already discussed a catalyst functions by providing a new pathway of lower activation energy. In homogeneous catalysis, it does so by forming an intermediate compound with one of the reactants. The highly reactive intermediate compound then reacts with the second reactant to yield the product, releasing the catalyst Let us illustrate it by taking the general reaction

$$A + B \xrightarrow{C} AB \qquad ...(1)$$

where C acts a catalyst. The reaction proceeds through the reactions:

$$\underset{\text{}}{A + C \rightarrow \underset{\text{Intermediate}}{AC}} \qquad ...(2)$$

$$AC + B \rightarrow AB + C \qquad ...(3)$$

The activation energies of the reactions (2) and (3) are lower than that of the reaction (1). Hence the involvement of the catalyst in the formation of the intermediate compound and its subsequent decomposition, accelerates the rate of the reaction (1) which was originally very slow.

Example : Catalytic oxidation of sulphur dioxide (SO_2) in the presence of nitric oxide (NO) as catalyst. (Chamber Process for Sulphuric acid)

$$2SO_2 + O_2 \xrightarrow{NO} 2SO_3$$

Mechanism

$$2NO + O_2 \rightarrow 2NO_2 \qquad \text{(Intermediate compound)}$$

$$NO_2 + SO_2 \rightarrow SO_3 + NO$$

Example : Formation of methylbenzene, $C_6H_5CH_3$ by reaction between benzene, C_6H_6, and methyl chloride, CH_3Cl, using aluminium chloride, $AlCl_3$, as catalyst (Friedel-Crafts reaction),

$$C_6H_6 + CH_3Cl \xrightarrow{AlCl_3} C_6H_5CH_3 + HCl$$

Mechanism

$$CH_3Cl + AlCl_3 \rightarrow [CH_3]^+ [AlCl_4]^-$$

Intermediate compound

$$C_6H_6 + [CH_3]^+ [AlCl_4]^- \rightarrow C_6H_5CH_3 + AlCl_3 + HCl$$

It may be noted that the actual isolation of intermediate compounds which would prove then existence is very difficult. As already stated, by their very nature they are unstable. In general, the intermediate compounds suggested as being formed are usually plausible rather than proved.

The Adsorption Theory

This theory explains the mechanism of a reaction between two gases catalysed by a solid (*Heterogeneous or Contact Catalysis*). Here the catalyst functions by adsorption of the reacting molecules on its surface.

Just like surface tension, the catalyst surface has unbalanced chemical bonds on it, The reactant gaseous molecules are adsorbed on the surface by these free bonds. This accelerates the rate of the reaction.

The distribution of free bonds on the catalyst surface is not uniform. These are crowded at the 'peaks', 'cracks' and 'corners' of the catalyst. The catalytic activity due to adsorption of reacting molecules is maximum at these spots. These are, therefore, referred to as the active centres. The active centres increase the rate of reaction not only by increasing the concentration of the reactants but they also activate the molecule adsorbed at two such centres by stretching it.

The adsorption theory explains catalytic activity

Metals in state of fine subdivision or colloidal form are rich in free valence bonds and hence they are more efficient catalysts than the metal in lumps.

PHOTOCHEMICAL REACTIONS

Ordinary reactions occur by absorption of heat energy from outside. The reacting molecules are energised and molecular collisions become effective. These bring about the reaction. The reactions which are caused by heat and in absence of light are called *thermal* or *dark reactions.*

On the other hand, some reactions proceed by absorption of light radiations. These belong to the visible and ultraviolet regions of the electromagnetic spectrum (2000 to 8000 A). The reactant molecules absorb photons of light and get excited. These excited molecules then produce the reaction.

A reaction which takes place by absorption of the visible and ultraviolet radiations is called a photochemical reaction.

The branch of chemistry which deals with the study of photochemical reactions is called *photochemistry.*

In photochemical reactions, the required energy is provided by the photons of visible or ultraviolet radiation.

In ordinary chemical reactions, the energy of activation is supplied by the chance collection in a molecule or a pair of molecules of a large amount of thermal energy. An alternative way in which the necessary activation energy can be acquired is through the absorption of quanta of visible or ultraviolet radiation. Reactions which follow as a result of energy so acquired are classified as *photochemical* reactions. With this description, photochemistry appears as a special branch of kinetics. In practice, the theory and the experimental arrangements used in photochemistry set it off as a rather special subject. The goal of these reaction studies, however, is still the elucidation of the mechanism of the reaction.

In photochemical reactions, energy is acquired by the molecules of one of the reactants as a result of the absorption of electromagnetic radiation. An early recognition by Einstein was that each molecule is activated by the absorption of one photon. This has led to the use of the Einstein unit for 1 mol of photons.

The energy of 1 mol, or Einstein, of photons of yellow light with a wavelength of 600 nm is 200 kJ. The energy of 1 mol, or 1 Einstein, of photons of near ultraviolet radiation, with wavelength 200 nm, is about 600 kJ. Ultraviolet, and in some cases visible, radiation can initiate photochemical reactions.

From these energy values and the bond energies deduced in the thermodynamic studies, the absorption of light in the visible or ultraviolet can be expected to be sufficient to break a chemical bond or at least to produce a high-energy reactive molecule.

The amount of the chemical reaction that occurs in a photochemical experiment is related to the amount of light absorbed.

The experimental arrangement for a photochemical experiment is indicated. The data obtained by using a setup that has been suitably calibrated allow the determination of the number of light quanta absorbed in a particular experiment. A photochemical reaction follows from the absorption of a quantum of radiation by a species in the reaction mixture. The initial consequence of this absorption of radiation is known as the primary process in the photochemical reaction.

The absorption of radiation might produce an excited electronic state which may correspond to an electronic form in which the electrons of a bond of the absorbing molecule no longer maintain a chemical bond. In such cases the absorption of radiation leads to the dissociation of the absorbing molecule.

Furthermore, the fragments obtained may be in their lowest or in some excited electronic state. In almost all photochemical studies, when a bond breaks as a result of the absorption of a quantum of radiation, a homolytic cleavage of a chemical bond occurs; *i.e.,* dissociation leads to the formation of free radicals. This leads us to expect that any chemical reactions that result will be initiated by these free radicals.

The absorption of radiation may lead to an excited, but bound, state. Such species have the necessary activation energy for many chemical reactions. The photochemical consequences of the formation of such a species depend on the lifetime of this species, or some other related high-energy species, compared with the average time between collisions of this species and a molecule with which it can react. A photochemical reaction can therefore be expected to occur if the excited state is a repulsive one and the free radicals that are formed do not immediately recombine. A reaction can also result if the lifetime of the bound excited state is long enough for a collision to occur which forms intermediate or product species.

One can determine the number of quanta absorbed and the number of molecules of reactant used or of a product formed in a given experiment. From these data the quantum yield, or quantum efficiency, defined as the ratio of the number of molecules reacting or produced to the number of quanta absorbed, is determined. Some reactions have quantum yields close to unity. The extremes, however, go from quantum yields of zero for molecules that absorb visible or ultraviolet radiation and show no photochemical reactions to quantum yields as high as 106 for chain reactions like that discussed for the reaction of H_2 and Cl_2.

PRIMARY AND SECONDARY REACTIONS

The overall photochemical reaction may consist of:

(a) A primary reaction

(b) Secondary reactions.

A primary reaction proceeds by absorption of radiation.

A secondary reaction is a thermal reaction which occurs subsequent to the primary reaction

For example, the decomposition of HBr occurs as follows: reaction.

$HBr + h\nu \rightarrow H + Br$	Primary reaction
$HBr + H \rightarrow H_2 + Br$	Secondary reaction
$Br + Br \rightarrow Br_2$	Secondary reaction
$2HBr + h\nu \rightarrow H_2 + Br_2$	Overall reaction

Evidently, the primary reaction only obeys the law of photochemical equivalence strictly. The secondary reactions have no concern with the law.

Quantum Yield (or Quantum Efficiency)

It has been shown that not always a photochemical reaction obeys the Einstein law. The number of molecules reacted or decomposed is often found to be markedly different from the number of quanta or photons of radiation absorbed in a given time.

The number of molecules reacted or formed per photon of light absorbed is termed Quantum yield. It is denoted by ϕ so that

$$\phi = \frac{\text{No. of molecules reacted or formed}}{\text{No. of photons absorbed}}$$

For a reaction that obeys strictly the Einstein law, one molecule decomposes per photon, the quantum yield $\phi = 1$. When two or more molecules are decomposed per photon, $\phi > 1$ and the reaction has a high quantum yield. If the number of molecules decomposed is less than one per photon, the reaction has a low quantum yield.

Cause of High Quantum Yield

When one photon decomposes or forms more than one molecule, the quantum yield $\phi > 1$ and is said to be high. The chief reasons for high quantum yield are :

Reactions subsequent to the Primary reaction. One photon absorbed in a primary reaction dissociates one molecule of the reactant. But the excited atoms that result may start a subsequent secondary reaction in which a further molecule is decomposed

$$AB + hv \rightarrow A + B \qquad \text{Primary}$$

$$AB + A \rightarrow A_2 + B \qquad \text{Secondary.}$$

Obviously, one photon of radiation has decomposed two molecules, one in the primary reaction and one in the secondary reaction. Hence the quantum yield of the overall reaction is 2.

A reaction chain forms many molecules per photon. When there are two or more reactants, a molecule of one of them absorbs a photon and dissociates (primary reaction). The excited atom that is produced starts a secondary reaction chain.

$$A_2 + hv \rightarrow 2A \qquad ...(1) \qquad \text{Primary}$$

$$A + B_2 \rightarrow AB + B \qquad ...(2) \qquad \text{Secondary}$$

$$B + A_2 \rightarrow AB + A \qquad ...(3) \qquad \text{reaction chain.}$$

It is noteworthy that A consumed in (2) is regenerated in (3). This reaction chain continues to form two molecules each time. Thus, the number of AB molecules formed in the overall reaction per photon is very large. Or that the quantum yield is extremely high.

Examples of High Quantum Yield

The above reasons of high quantum yield are illustrated by citing examples as below :

(i) *Decomposition of HI* : The decomposition of hydrogen iodide is brought about by toe absorption of light of less than 4000Å. In the primary reaction, a molecule of hydrogen iodide absorbs a photon and dissociates to produce H and I. This is followed by secondary steps as shown below:

$$HI + hv \rightarrow H + I \qquad ...(1) \qquad \text{Primary}$$

$$H + HI \rightarrow H_2 + I \qquad ...(2) \qquad \text{Secondary}$$

$$I + I \rightarrow I_2 \qquad ...(3)$$

$$2HI + h\nu \rightarrow H_2 + I_2 \qquad \text{Overall reaction}$$

In the overall reaction, two molecules of hydrogen iodide are decomposed for one photon (hv) of light absorbed. Thus the quantum yield is 2.

PHOTOSENSITIZED REACTION

A species which can both absorb and transfer radiant energy for activation of the reactant molecule is called a photosensitized. The reaction so caused is called a photosensitized reaction. Now let us study about some of the examples of photosensitized Reaction.

Reaction Between H_2 and O_2 : This reaction is photosensitized by mercury vapour. The product is hydrogen peroxide, H_2O_2.

$$Hg + h\nu \rightarrow Hg \qquad \text{Primary absorption}$$

$$Hg + H_2 \rightarrow 2H + Hg \qquad \text{Energy transfer}$$

$$H + O_2 \rightarrow HO_2$$

$$HO_2 + HO_2 \rightarrow H_2O_2 + O_2 \qquad \text{Reaction}$$

Hydrogen peroxide may decompose to form water, H_2O.

Reaction Between H_2 and CO : Mercury vapour is used as photosensitizer. The product is formaldehyde, HCHO.

$$Hg + h\nu \rightarrow Hg^+ \qquad \text{Primary absorption}$$

$$Hg + H_2 \rightarrow 2H + Hg \qquad \text{Energy transfer}$$

$$H + CO \rightarrow HCO$$

$$HCO + H_2 \rightarrow HCHO + H \qquad \text{Reaction}$$

$$2HCO \rightarrow HCHO + CO.$$

Some glyoxal, CHO—CHO, is also formed by dimerization of formyl radicals, HCO. Some of the photophysical processes also cause chemical changes. Now let us deal with such photophysical processes.

Fluorescence

Certain molecules (or atoms) when exposed to light radiation of short wavelength (high frequency), emit light of longer wavelength. The process is called fluorescence and the substance that exhibits fluorescence is called fluorescent substance. Fluorescence stops as soon as the incident radiation is cut off.

Examples :

(a) a solution of quinine sulphate on exposure to visible light, exhibits blue fluorescence.

(b) a solution of chlorophyll in ether shows blood red fluorescence.

Explanation : When a molecule absorbs high energy radiation, it is excited to higher energy states. Then it emits excess energy through several transitions to the ground state. Thus the excited molecule emits light of longer frequency. The colour of fluorescence depends on the wavelength of light emitted.

Phosphorescence

When a substance absorbs radiation of high frequency and emits light even after the incident radiation is cut off, the process is called phosphorescence. The substance which shows phosphorescence is called phosphorescent substance. Phosphorescence is chiefly caused by ultraviolet and visible light. It is generally shown by solids.

Examples : Sulphates of calcium, barium and strontium exhibit phosphorescence.

Fluorescence in boric acid shows phosphorescence in the blue region at 5700Å wavelength.

Explanation : As in fluorescence, a molecule absorbs light radiation and gets excited. While returning to the ground state, it emits light energy of longer wavelength. In doing so the excited molecule passes from one series of electronic states to another and gets trapped. This shows the emission of light which persists even after the removal of light source. Thus phosphorescence could be designated as delayed fluorescence.

$$A + h\nu \rightarrow A^{+} \xrightarrow{\text{slow}} h\nu'.$$

Chemiluminescence

Some chemical reactions are accompanied by the emission of visible light at ordinary temperature. *The emission of light as a result of chemical action is called chemiluminescence.* The reaction is referred to as a *chemiluminescent reaction.* Such a reaction is the reverse of a photochemical reaction which proceeds by absorption of light. The light emitted in a chemiluminescent reaction is also called '*cold light*' because it is produced at ordinary temperature. The glow of fireflies due to the aerial oxidation of *luciferon* (a protein) in the presence of enzyme *luciferase.* The oxidation of 5-*animophthtalic* cyclic hydrazide (*luminol*) by hydrogen peroxide in alkaline solution, producing bright green light.

5

Types of Solids in Chemistry

INTRODUCTION

In both gases and liquids, atoms, ions and molecules continually move. They translate randomly as well as rotate and vibrate. This determines the ability of gases and liquids to flow. In solids, atoms, ions and molecules are held together by relatively strong chemical forces—ionic bond, covalent bond, or by intermolecular van der Waals' forces. They do not translate although they vibrate to some extent in their fixed positions. This explains why solids are rigid and have definite shape.

Gases and liquids can flow and take up the shape of their container. Solids are rigid and lack the ability to flow.

Solids can be broadly classified into two types:

(a) Crystalline solids; also called true solids

(b) Amorphous solids

A *crystalline solid* exists as small crystals, each crystal having a characteristic geometrical shape.

In a crystal, the atoms, molecules or ions are arranged in a regular, repeating three-dimensional pattern called the *crystal lattice*. Sugar and salt are crystalline solids.

An *amorphous solid* (*Gr amorphos* = no form) has atoms, molecules or ions arranged at random and lacks the ordered crystalline lattice. Examples are rubber, plastics and glass. In their disordered structure, amorphous solids resemble liquids. Thus glasses are to be regarded as supercooled or highly viscous liquids. The liquid nature of glass is sometimes apparent in very old window panes that have become slightly thicker at the bottom due to gradual downward flow.

Isotropy and Anisotropy

Amorphous substances are said to be isotropic because they exhibit the same value of any property in all directions. Thus refractive index, thermal and electrical conductivities, coefficient of thermal expansion in amorphous solids are independent of the direction along which they are measured.

Crystalline substances, on the other hand, are anisotropic and the magnitude of a physical property varies with directions. For example, in a crystal of silver iodide, the coefficient of thermal expansion is positive in one direction and negative in the other. Similarly, velocity of light in a crystal may vary with direction in which it is measured. Thus a ray of light passing through a Nicol prism splits up into two components, each travelling with different velocity (double refraction).

In amorphous substances, as in liquids, the arrangement of particles is random and disordered. Therefore all directions are equivalent and properties are independent of directions.

On the other hand, the particles in a crystal are arranged and well ordered. Thus, the arrangement of particles may be different in different directions. The external shape is called the *habit* of the crystal. The plane surfaces of the crystal are called *faces*. The angles between the faces are referred to as the *interfacial angles*. The interfacial angles for a given crystalline substance are always the same.

The constancy of interfacial angles is an essential characteristic of crystalline solids.

STRUCTURE OF A CRYSTAL

The particles in crystals are arranged in regular patterns that extend in all directions. The overall arrangement of particles in a crystal is called the *Crystal lattice*, *Space lattice* or *Simply lattice*.

To describe the structure of a crystal it is convenient to view it as being made of a large number of basic units. The simple basic unit or the building block of the crystal lattice is called the *Unit cell*.

The crystal lattice of a substance is depicted by showing the position of particles (*structural units*) in space. These positions are represented by bold dots (or circles) and are referred to as *lattice points* or *lattice sites*. The overall shape and structure of a crystal system is governed by that of the unit cell of which it is composed.

A unit cell has one atom or ion at each corner of the lattice. Also, there may be atoms or ions in faces and interior of the cell. A cell with an interior point is called the *body centered cell.* A cell which does not contain any interior points is known as the primitive cell. That is *a primitive cell is a regular three-dimensional unit cell with atoms or ions located at its corners only.*

Parameters of the Unit Cells

In 1850, August Bravis, a French mathematician observed that the crystal lattice of substances may be categorised into seven types. These are called *Bravis lattices* and the corresponding unit cells are referred to as *Bravis unit cells.* The unit cells may be characterised by the following *parameters.*

Cubic unit cells are called simplest unit cells. They got importance for two reasons. First, a number of ionic solids and metals, have crystal lattices comprising cubic unit cells. Secondly, it is easy to make calculations with these cells because, they have equal sides and cells angles are 90°.

Now let us try to understand about the parameters seven unit cells, with the diagram.

By studying the following table we can learn about the relative axial length and angles of a crystal system.

Table 1

Crystal System	Relative Axial Length	Angles
Cubic (isometric)	$a = b = c$	$\alpha = \beta = \gamma = 90°$
Tetragonal	$a = b \neq c$	$\alpha = \beta = \gamma = 90°$
Orthorhombic	$a \neq b \neq c$	$\alpha = \beta = \gamma = 90°$
Rhombohedral (trigonal)	$a = b = c$	$\alpha = \beta = \gamma \neq 90°$
Hexagonal	$a \neq b \neq c$	$\alpha = \beta = 90°, \gamma - 120°$
Monoclinic	$a = b \neq c$	$\alpha = \beta = 90°, \gamma \neq 90°$
Triclinic	$a \neq b \neq c$	$\alpha \neq \beta \neq \gamma \neq 90°$

BRAGG'S EQUATION

In 1913 the father-and-son, W.L. Bragg and W.H, Bragg worked out a mathematical relation to determine interatomic distances from

X-ray diffraction patterns. This relation is called the *Bragg equation.* According to them,

(1) *The X-ray diffracted from atoms in crystal planes obey the laws of reflection.*

(2) *The two rays reflected by successive planes will be in phase if the extra distance travelled by the second ray is an integral number of wavelengths.*

Derivation of Bragg Equation

Here two successive atomic planes of the crystal are shown separated by a distance d. Let the X-rays of wavelength A, strike the first plane at an angle 9. Some of the rays will be reflected at the same angle. Some of the rays will penetrate and get reflected from the second plane. These rays will reinforce those reflected from the first plane if the extra distance travelled by them (CB + BD) is equal to integral number, n, of wavelengths. That is,

$$n\lambda = CB + BD \qquad ...(i)$$

Geometry shows that

$$CB = BD = AB \sin \theta \qquad ...(ii)$$

From (i) and (ii) it follows that

$$n\lambda = 2AB \sin \theta$$

or

$$n\lambda = 2d \sin \theta.$$

This is known as the *Bragg equation.* The reflection corresponding to n = 1 (for a given series of planes) is called the first order reflection. The reflection corresponding to n = 2 is the second order reflection and so on.

Bragg equation is used chiefly for determination of the spacing between the crystal planes. For X-rays of specific wave length, the angle θ can be measured with the help of *Bragg X-ray spectrometer*. The interplanar distance can then be calculated with the help of Bragg equation.

The study of crystal structure with the help of x-rays is called x-ray crystallography. The diffraction pattern produced by crystals can be understood.

Bragg's Equation

When a beam of X-rays is allowed to fall on a crystal, a large number of images of different intensities are formed. If the diffracted waves are in the same phase, they reinforce each other and a series of bright spots are produced on a photographic plate placed in their path. On the other hand, if the diffracted waves are out of phase, dark spots are caused on the photographic plate. From the overall diffraction patterns produced by a crystal, we can arrive at the detailed information regarding the position of particles in the crystal.

MILLER INDICES

The directions of the special planes that pass through many lattice points can be described by Miller indices.

The law of rational indices states that the intercepts of the planes of the various faces of a crystal on a suitable set of axes can be expressed by small integral multiples of three unit distances.

By studying the following diagram we can understand, how planes that pass through, relatively large numbers of lattice points.

The empirical law describing the arrangement of the faces of a crystal is understandable if the internal structure of the crystal corresponds to an array of points that constitute a lattice. The important planes of the crystal then correspond to the planes that can be drawn to include a relatively large number of lattice points. The orientation of the faces of a crystal can then be compared with the orientation and spacing of the planes in the 14 possible Bravais lattices. In this way the internal structure of the crystal can be identified with a particular lattice.

The relative orientation of crystal or lattice planes is of great importance in crystal-structure analysis, and a convenient method for describing these planes is needed. The important planes of a lattice, in terms of intercepts that are multiples of the unit-cell dimensions. Since crystal planes are similarly oriented, they can be similarly described. In this way Weiss indices are used, and planes are described by their relative intercepts on the x, y, and z axes as a: b: c, or a: 2b: ∞ c, and so forth.

Much more convenient, particularly in the analysis of diffraction data, are sets of numbers called Miller indices. These are obtained by taking the reciprocals of the coefficients of the Weiss indices. These three reciprocals are then cleared of fractions and reduced to the smallest set of integers.

The plane a: b: c becomes a (111) plane; the a : 2b : c plane becomes a (212) plane; the a:b : ∞ c plane becomes a (110) plane. The Miller indices, referred to in general by (hkl), describe the relative directions of the crystal planes. Information on the coordinate system and the values of the three unit distances is given separately.

A feeling for these indices can soon be acquired if one remembers their reciprocal nature. A low first number means an intercept at a large distance on the x axis; a low second number means a large intercept on the y axis; and so forth. A plane parallel to an axis now has a zero term corresponding to the reciprocal of its intercept on that axis.

Miller indices provide a convenient way of expressing a particular plane, *i.e.,* a particular direction in a crystal. It should be pointed out that as far as direction is concerned, which is sometimes all that is important, planes (220) and (110) would be the same, and the latter notation would be used.

THE DIFFRACTION SPOTS

Nowadays, the precession technique is widely used to study the crystal structure. In this technique the crystal is moved in a precession motion in the x-ray beam. The precession method spreads out all the diffraction spots having a selected value one of the indices, h,k or *l* over the film. The values of the remaining two indices can be determined. Thus, the diffraction spots that result can be assigned all three indices, h, k and *l*. Then the shape and complete dimensions of the unit cell can be deduced.

Consider a crystal mounted in an x-ray beam and then tilted relative to the beam so that the $l = 0$ diffractions, *i.e.,* all those which would be indexed as hk0 can be produced. If the crystal has $a = b$, that is, is cubic or tetragonal, and it is positioned diffraction from the (100) could occur and would form the indicated spot.

Now, suppose that the crystal is processed so that the x-ray beam impinges on the crystal. A 010 diffraction will be recorded, as shown.

In a similar way, if the precession is continued, 100 and 010 diffractions, the overbar implying minus values, will form spots at the bottom and at the side, to form the four-spot pattern produced by this set of planes. Many more diffractions would be observed with the complete range of precession implied. The additional spots can also be related to crystal planes. The (110) planes, for example, lead to spots along the diagonals of the film. Most important is the recognition that any *diffraction spot will appear on the film in a direction from the film center that is perpendicular to the set of planes that produces the diffraction.*

When rows of spots can be recognized in a precession photograph, they can be associated with sets of planes that have a common degree of tilt relative to a given direction. Thus, the 0k0 diffractions are found to be in an equatorial-layer-line array of spots. The planes that produce these spots have no tilt relative to the a axis. Other lines of spots due to planes with various amounts of tilt relative to this axis are 1k0, 2k0, 3k0, and so forth. Similar indexing of spots by amounts of tilt relative to the b axis gives the arrays h10, h20, ... Complete indices of the spots on the interlocking layer lines is then achieved. Thus the diffraction spots occurs in a direction perpendicular to the plane directions.

So far we have not used the information provided by the distance of the all fraction spots from the center of the film. The precession method again introduces a mechanically complicated procedure that leads to an easy interpretation of these distances. Thus, the film obtained in the precession method is a display of the orientation of the crystal planes and their spacing. The orientation depends on the fact that the direction of the diffraction spot from the center of the film is the *direction* perpendicular to the plane. The spacing results because the *distance* of the spot from the film center is proportional to the reciprocal of the plane spacing. The precession-method film is said to exhibit the *reciprocal lattice* of the crystal.

SINGLE CRYSTAL DIFFRACTION AND POWDER DIFFRACTION

For a crystal arranged and subjected to a rotation about one of the axes, the a axis, the various crystal planes parallel to the a axis will satisfy Bragg's law and will produce reflections along the "equatorial" plane shown by the Miller index h = 0.

Planes tilted with respect to the a axis will produce reflection spots along *layer lines* above and below the equatorial plane. The degree of tilt, the index h, and the layer lines.

Arrangement of the Bragg expression to

$$\lambda = 2 \sin \theta = 2d_{hkl} \sin \theta,$$

shows that the higher-order reflections from planes with a spacing d can be treated as if they were due to first-order reflections from planes with spacing $d_{hkl} = d/n$. Thus each reflection can be labeled with a value of hkl corresponding to the Miller indices (hk*l*) of the planes that give rise to the reflection. A spacing d_{hkl} can be associated with each hkl reflection. A reflection that is labeled, or indexed, as 200, for example, would have associated with it a spacing half that of the (100) planes. This procedure is more satisfactory than treating the reflection as a second-order reflection from planes with the (100) spacing. Instead, therefore, of recognizing reflections of various orders of the (0k*l*) planes in the equatorial layer line, for example, we label, or index, the reflections as 010, 020, 030 and 011, 022, 033, and so forth. The higher-order indices correspond to plane spacings that are fractions of the lattice point spacings.

Similar indexing of the reflections in the other layer lines can be done, and all these reflections may need to be considered to deduce the nature of the internal structure of the crystal.

Powder Diffraction

The crystals in a powdered sample present all possible orientations to the x-ray beam. The diffraction obtained will be just like that which would result from mounting a single crystal and turning it through all possible angles. For each crystal plane there will be some one angle at which the Bragg law will be satisfied, and some of the crystals will have this orientation; Since there are quite a few crystal planes with a fairly high density of particles, there will be reflections from each of these, and the pattern will show scattering at a large number of angles.

For relatively simple crystals, the powder pattern can be used in the study of the crystal structure. For more complex systems, particularly those of low symmetry, many planes in the crystal will happen to have equal or nearly equal spacing, and even if these planes have different directions in the crystal, the powder method will superimpose the reflections from these planes. It is difficult to associate particular diffractions with particular lattice planes.

The most widespread application of the powder method, which is far more easily used, than the single-crystal method is in the qualitative analysis of solid samples.

The spacings between crystal planes, and therefore the position of the line of a powder photograph are characteristic of a given crystal. The components of a solid mixture can often be rapidly identified from a powder pattern.

ARRANGEMENT OF IONS IN THE UNIT CELL

X-ray diffraction techniques have given us a fund of precise, reliable data on the dimensions of atomic, ionic, and molecular systems. Let us see some of the generalizations that can be drawn from such data. We begin, in this section, by considering the simple ions of salt crystals. Such ions are represented as spheres. The simplest and very useful approach treats the radius of a given ion as a fixed quantity independent of the type and number of ions that surround it in a particular crystal.

Consideration of the contributions to the energy of an ionic crystal suggests that normally anions and cations are in contact with each other. If this is the case, the sum of the anion and cation radii can be deduced from the unit-cell dimensions. Examples of the necessary calculations are given for both AB- and AB_2-type crystals.

A number of methods have been suggested for obtaining individual ionic radii from these relatively easily obtained sums. Perhaps the simplest depends on the assumption that the crystal with the largest anion I^- and the smallest cation Li^+ will present the situation. In this case the cation is so small that there is anion-anion contact and the radius of the anion, the iodide ion, can be obtained from the unit-cell dimensions, as suggested in the figure. Once a value for the radius of one ion is obtained by this or other methods not dealt with here, a table of ionic radii can be constructed from the sums of ionic radii deduced from other crystal structures. Generally the ionic radius is greater as, the co-ordination number of the ion becomes greater.

CLASSIFICATION OF CRYSTALS

Depending on the basis of bonds crystals can be classified as follows:

1. Ionic Crystals.
2. Molecular Crystals.
3. Network Covalent Crystals.
4. Metallic Crystals.

1. Ionic Crystals

In an ionic crystal the lattice is made of positive and negative ions. These are held together by ionic bonds - the strong electrostatic attractions between oppositely charged ions. Consequently, the cations and anions attract one another and pack together in an arrangement so that the attractive forces maximise. Each ion is surrounded by neighbours of opposite charge and there are no separate molecules. Since the ions are fixed in their lattice sites, typical ionic solids are hard and rigid with high melting points. In spite of their hardness, ionic solids are brittle. They shatter easily by hammering. By hammering, a layer of ions slips away from their oppositely charged neighbours and brings them closer to ions of like charge. The increase of electrostatic repulsions along the displaced plane causes the crystal to break.

Ionic solids are non-conducting because the ions are in fixed positions. However in the fused state the ions are allowed freedom of movement. So that, it becomes possible for them to conduct electricity.

Now let us study about the crystal lattice of sodium chloride.

Each sodium ion is surrounded by six chloride

ions and each chloride ion is clustered by six sodium ions. The co-ordination number for this crystal lattice is six as required by simple cubic type. In this cubic system, the planes can be passed through the atoms

having Miller indices (100), (110) or (111) and the relative spacings for the unit cell of a face-centred cubic lattice are $\frac{a}{2}, \frac{a}{2\sqrt{2}}$ and $\frac{a}{\sqrt{3}}$ while it is $a : \frac{a}{\sqrt{2}} : \frac{a}{\sqrt{3}}$ for simple cubic and $\frac{a}{2} : \frac{a}{\sqrt{2}} : \frac{a}{2\sqrt{3}}$ for body-centred cubic lattice. For face-centred cubic lattice,

$$d_{100} : d_{110} : d_{111} = \frac{a}{2} : \frac{a}{2\sqrt{2}} : \frac{a}{\sqrt{3}}$$

$$= 1 : 0.707 : 1.154 \qquad \text{...(i)}$$

In the case of sodium chloride the *first order* reflections from (100), (110) and (111) faces using K line from palladium anti-cathode are 5.9°, 8.4° and 5.2° respectively. From Bragg equation $n\lambda = 2d \sin\theta$, we have

$$d = \frac{n\lambda}{2 \sin\theta}$$

Since n = 1 and X is the same in each case, the ratio of the spacings parallel to the three principal planes are

$$d_{100} : d_{110} : d_{111} = \frac{1}{\sin 5.9^\circ} : \frac{1}{\sin 8.4^\circ} : \frac{1}{\sin 5.2^\circ}$$

$$= 9.731 : 6.844 : 11.04$$

$$= 1 : 0.704 : 1.136 \qquad \text{...(ii)}$$

This ratio is almost identical with the ratio required in the case of a face-centred cubic lattice as shown in (i).

X-ray diffraction studies reveal that unlike sodium chloride, potassium chloride has a simple cubic lattice. The edge length of the unit cell is 3.1465Å.

2. Molecular Crystals

In Molecular crystals, molecules are held together by van der Waal's forces. When this type of crystal melts it is only the weak van der Waal's forces that must be overcome. Therefore molecular solids have low melting points. Most organic substances are molecular solids.

Crystals Lattice of Dry CO_2

Dry ice or frozen carbon dioxide is the best example of a molecular solid. Then van der Waal's forces holding the CO_2 molecules together are weak enough so that dry ice passes from solid state to gaseous state at 78°C.

3. Network Covalent Crystals

In this type of crystals atoms occupy the lattice sites. These atoms are bonded to one another by covalent bonds. The atoms interlocked by a network of covalent bonds produce a crystal which is considered to be a single giant molecule. Such a solid is called a *network covalent solid* or *simply covalent solid.* Since the atoms are bound by strong covalent bonds, these crystals are very hard and have very high melting points.

4. Metallic Crystals

The crystals of metals consist of atoms present at the lattice sites. The atoms are arranged in different patterns, often in layers placed one above the other. The atoms in a metal crystal are viewed to be held together by a metallic bond. The valence electrons of the metal atoms arc considered to be delocalised leaving positive metal ions. The freed electrons move throughout the vacant spaces between the ions. The electrostatic attractions between the metal ions and the electron cloud constitute the metallic bond. Thus a metal crystal may be described as having positive ions at the lattice positions surrounded by mobile electrons throughout the crystal.

THE COMMON CRYSTAL DEFECTS

A perfect crystal is one in which all the atoms or ions are lined up in a precise geometric pattern. But crystals are never actually perfect. The *real crystals* that we find in nature or prepare in the laboratory always contain imperfections in the formation of the crystal lattice. These crystal defects can profoundly affect the physical and chemical properties of a solid.

The common crystal defects are:

(a) Vacancy Defect

(b) Interstitial Defect

(c) Impurity Defect.

These defects pertaining to lattice sites or points are called *Point defects*.

(a) Vacancy Defect

When a crystal site is rendered vacant by removal of a structural unit in the lattice, the defect is referred to as the *vacancy defect*. In an ionic crystal, a cation and an ion may leave the lattice to cause two vacancies. Such a defect which involves a cation and an anion vacancy in the crystal lattice is called a *Schottky defect* This defect is found in the crystals of sodium chloride and cesium chloride (CsCl).

(b) Interstitial Defect

Here, an ion leaves its regular site to occupy a position in the space between the lattice sites (interstitial position). This causes a defect known as Interstitial defect or Frenkel defect. Ordinarily the cation moves as it is smaller than the anion and can easily fit into the vacant spaces in the lattice. Thus, in Ag*Cl* crystal, Ag^+ ion occupies an interstitial position leaving a vacancy (or hole) at the original site.

(c) Impurity Defects

These defects arise due to the corporation of foreign atoms or ions in regular lattice sites or interstitial sites.

NEUTRON DIFFRACTION

An x-ray beam is scattered primarily as a result of interaction with the electrons that surround each atom of an ion or molecule. Thus, the atomic scattering factors are approximately proportional to atomic numbers. As a result, it is difficult to locate low-atomic-number atoms in the presence of high-atomic-number atoms. Thus the position of hydrogen atoms generally cannot be deduced from x-ray diffraction studies. (A positive attitude to this difficulty is the recognition that the positions of hydrogen atoms are of little consequence and thus do not add to the complexity of many already complex structure determinations.) Within a factor of 3 or 4, all nuclei scatter neutrons to the same extent. Thus hydrogen-atom nuclei or, more conveniently, deuterium nuclei can be located in the presence of heavier atoms.

Neutron diffraction is also distinguished by the noticeable role of the magnetic moment of the neutron when diffraction occurs from crystals with ordered atomic magnetic moments. Neutron diffraction can be used to study the orientation of atomic magnetic moments in ferromagnetic and antiferromagnetic crystals. Neutron diffraction is thus a specialized adjunct to x-ray diffraction.

A neutron beam is usually formed from the thermal neutrons of a nuclear reactor. A velocity-selector device augmented by crystal diffraction provides a monochromatic beam. The wavelength associated with the neutron beam can be calculated from the de Broglie relation $\lambda = h/(m\upsilon)$. The momentum term mv is obtained for thermal neutrons by setting

$$\frac{1}{2}m\upsilon^2 = \frac{3}{2}kT$$

and rearranging to get

$$m\upsilon = \sqrt{3mkT}$$

With appropriate numerical values, including 1.675×10^{-27} kg for the mass of a neutron, we obtain

$$\lambda = 1.46 \times 10^{-10} \text{ m} = 146 \text{ pm}$$

A beam of thermal neutrons therefore has a wavelength suitable for diffraction studies of crystals. A diffraction pattern like those obtained for x-ray diffraction, except that a longer-wavelength beam is used, is obtained.

Typically crystals whose structures are known, except for the positions of hydrogen atoms, are studied. The Fourier-transform procedure is used so that the positions of these remaining nuclei are found. The smallness of the nuclei would lead to sharp peaks on a nuclear-density map, but thermal motion produces some spreading of these peaks.

ULTRAMICROSCOPE

The light scattering is the phenomenon on which the ultramicroscope is based.

The scattering of light by solutions of polymers or macromolecules gives information on the shape of these molecules.

Some information about the size and shape of colloidal and "macromolecular" particles can be obtained from measurements of the visible light that is scattered by "solutions" of these materials. Light scattering is familiar as the *Tyndall effect*, the noticeable beam of light seen when the beam passes through smoky or misty air or through a milky liquid. The sample is observed through a microscope at right angles to the direction of the entering light beam. Each colloid particle, larger than about 1-nm diameter in very favourable cases, will produce an observable point of scattered light. The individual particles can then be counted, and if the microscope focuses on a definite, known volume of solution, the number of particles per unit volume can be determined. Such data, along with the measurable mass of macromolecule material per unit volume, lead to a value of the average mass of the individual particles.

None of the details of the particles can be observed. They merely act as scattering centers, and one observes points of light. Furthermore, unless the refractive index of the colloid particle is very different from that of the solvent, the scattered light is too weak to be seen. The similarity of the refractive index of most macro-molecules to the medium in which they are dispersed means that little scattered light is given off. The method is therefore most applicable to inorganic colloids.

The amount of scattered light and the direction in which this light is scattered can be related to the size and shape of the scattering particles.

THE LIQUID CRYSTALS

Some organic solids having long rod-like molecules do not melt to give the liquid substance directly. They, instead, pass through an intermediate state called the *liquid crystal state*, often referred to as the *liquid crystal*. Thus the liquid crystal state is intermediate between the liquid state

The liquid crystals have a structure between that of a liquid and that of a crystalline solid. In a liquid the molecules have a random arrangement and they are able to move past each other. In a solid crystal the molecules have an ordered arrangement and are in fixed positions. In a liquid crystal, however, molecules are arranged parallel to each other and can flow like a liquid. *Thus, the liquid crystals have the fluidity of a liquid and optical properties of solid crystals.*

Liquid Crystal Types

According to their molecular arrangement, the liquid crystals are classified into three types.

Nematic Liquid Crystals

They have molecules parallel to each other like soda straws but they are free to slide or roll individually.

Smectic Liquid Crystals

The molecules in this type of crystal are also parallel but these are arranged in layers. The layers can slide past each other.

Chloesteric Liquid Crystals

As in nematic crystals, in this type of crystal the molecules are parallel but arranged in layers. The molecules in successive layers are slightly rotated with respect to the layers above and below so as to form a spiral structure. On account of their remarkable optical and electrical properties, liquid crystals find several practical applications. Two of these are listed below.

When a thin layer of nematic liquid crystal is placed between two electrodes and an electric field is applied, the polar molecules are pulled out of alignment. This causes the crystal to be opaque. Transparency returns when electrical signal is removed. This property is used in the number displays of digital watches, electronic calculators, and other instruments.

FIRST LAW OF THERMODYNAMICS

The study of the flow of heat or any other form of energy into or out of a system as it undergoes a physical or chemical transformation, is called Thermodynamics.

In studying and evaluating the now of energy into or out of a system it will be useful to consider changes in certain properties of the system. These properties include temperature, pressure, volume and concentration of the system. Measuring the changes in these properties from the initial state to the final state, can provide information concerning changes in energy and related quantities such as heat and work.

Thermodynamics is a powerful method for studying chemical phenomena. It can be developed quite independently of the atomic and molecular theory of the preceding chapters, and it can be applied to systems of any complexity. The name *thermodynamics* suggests the origin of the subject in the conversion of thermal energy to mechanical energy in the engines that drove the industrial revolution. In modern chemistry we continue to be concerned with various types of energy and with the energy transfor-mations that occur in chemical systems. You will see that thermodynamics provides us with a remarkably useful energy book keeping system.

In addition, we use thermodynamic quantities that help us to understand the equilibrium state of chemical systems. This important aspect of many chemical studies is a modern counterpart of the early attempts to harness the tendency for temperatures to become uniform and a thermal equilibrium state to be reached. Much of our studies of thermodynamics in subsequent chapters is directed to associating energies and energy-related properties with the equilibrium state of chemical systems.

Thermodynamics is a logical subject of great elegance. Three concise statements, the three laws of thermodynamics, sum up our experiences with energy and natural processes. From these statements logical deductions are then drawn that bear on almost every aspect of chemistry. The study of thermodynamics is based on three broad generalisations derived from well established experimental results. These generalisations are known as *the First, Second and Third Law of Thermodynamics*. These laws have stood the test of time and are independent of any theory of the atomic or molecular structure.

Scope of Thermodynamics

1. Most of the important laws of Physical Chemistry, including the van't Hoff law of lowering of vapour pressure. Phase Rule and the Distribution Law, can be derived from the laws of thermo-dynamics.
2. It tells whether a particular physical or chemical change can occur under a given set of conditions of temperature, pressure and concentration.
3. It also helps in predicting how far a physical or chemical change can proceed, until the equilibrium conditions are established.

THE ENERGY OF A CHEMICAL SYSTEM

The energy of a chemical system depends on its volume.

In almost all the chemical examples considered in Part Two, the mechanical surroundings gain or lose energy only as a result of the expansion or contraction of the system. So that we can deal with processes in which the volume changes, we now develop a convenient expression for the mechanical-energy changes that this volume change produces.

The energy that can be transferred to the mechanical surroundings depends on the force that the connection to the mechanical system exerts on the piston shaft and the distance through which this force acts. If f is the force exerted on the shaft of the piston and dl is the incremental distance that the shaft and the piston move, the change in the energy of the mechanical surroundings when the volume of the system changes is given by the expression

$$dU_{mech} = fdl \quad ...(1)$$

This relation can be put in a form that is more useful when chemical systems are dealt with. If the piston area is A, the pressure corresponding to the force f is

P = f/A. We can write/= PA and change eq. (1) to

$$dU_{mech} = PAdl \quad ...(2)$$

When A dl is recognized as the change of volume dV of the system, this becomes the often useful relation

$$dU_{mech} = PdV \quad ...(3)$$

Notice that if the volume of the system increases, the piston is driven back and the energy of the mechanical surroundings increases. If the volume of the system decreases, the energy of the mechanical surroundings decreases. We have, in eq. (3), the basis for calculating these energy changes.

Often in chemical reactions the energy change due to the change in the volume of the system is small compared to the total energy change of a chemical reaction. We need, however, to take care of this "annoying" detail so that our treatment of the energies of chemical processes is on a firm foundation.

NON-EQUILIBRIUM STATES

A system in which the state variables have constant values throughout the system is said to be in a state of thermodynamic equilibrium.

Suppose we have a gas confined in a cylinder that has a frictionless piston. If the piston is stationary, the state of the gas can be specified by giving the values of pressure and volume. The system is then in a state of equilibrium.

A system in Which the state variables have different values in different parts of the system is said to be in a non-equilibrium state.

If the gas contained in a cylinder, is compressed very rapidly by moving down the piston, it passes through states in which pressure and temperature cannot be specified, since these properties vary throughout the gas. The gas near the piston is compressed and heated and that at the far end of the cylinder is not The gas then would be said to be in non-equilibrium state.

Thermodynamics is concerned only with equilibrium states.

The Criteria for Equilibrium

(1) *The temperature of the system must be uniform and must be the same as the temperature of the surroundings (thermal equilibrium).*

(2) *The mechanical properties must be uniform throughout the system (mechanical equilibrium).* That is, no mechanical work is done by one part of the system on any other part of the system.

(3) *The chemical composition of the system must be uniform with 10 net chemical change (chemical equilibrium).* If the system is heterogeneous, the state variables of each phase remain constant in each phase.

When a thermodynamic system changes from one state to another, the operation is called a Process. These processes involve the change of conditions (temperature, pressure and volume).

The various types of thermodynamic processes are:

(1) Isothermal Processes

Those processes in which the temperature remains fixed, are termed isothermal processes. This is often achieved by placing the system in a thermostat (a constant temperature bath).

For an isothermal process $dT = 0$.

(2) Adiabatic Processes

Those processes in which no heat can How into or out of the system, are called adiabatic processes. Adiabatic conditions can be approached. By carrying the process in an insulated container such as 'merinos' bottle.

High vacuum and highly polished surfaces help to achieve thermal insulation.

For an adiabatic process dq = 0.

(3) Isobaric Processes

Those processes which take place at constant pressure are called isobaric processes. For example, heating of water to its boiling point and its vaporisation take place at the same atmospheric pressure. These changes are, therefore, designated as isobaric processes and are said to take place sobarically.

For an isobaric process dp = 0.

(4) Isochoric Processes

Those processes in which the volume remains constant are known as isochoric processes.

The heating of a substance in a non-expanding chamber is an example of isochoric process.

For isochoric processes dV = 0.

(5) Cyclic Process

When a system in a given state goes through a number of different processes and finally returns to its initial state, the overall process is called a cycle or cyclic process.

For a cyclic processes 0, dH = 0.

ISOTHERMAL EXPANSION OF A GAS

The isothermal expansion of an ideal gas may be carried either by the *reversible process* or *irreversible process* as stated above.

The reversible expansion is the pressure is falling as the volume increases. The reversible work done by the gas in given by the expression

$$-w_{rev} = \int_{V_1}^{V_2} PdV$$

which is represented by the shaded area.

If the expansion is performed irreversibly by suddenly reducing the external pressure to the final pressure P_2, the irreversible work is given by

$$-w_{irr} = P_2 (V_2 - V_1)$$

In both the processes, the state of the system has changed from A to B but the work done is much less in the irreversible expansion than in the reversible expansion. *Thus, mechanical work is not a state function as it depends on the path by which the process is performed rather than on the initial and final states. It is a path function.*

It is also important to note that *the work done in the reversible expansion of a gas is the maximum work that can be done by a system (gas) in expansion between the same initial (A) and final state (B).* This is proved as follows:

We know that the work always depends on the external pressure, P_{ext}; the larger the P_{ext} the more work is done by the gas. But the P_{ext} on the gas cannot be more than the pressure of the gas, P_{gas}, or a compression will take place. Thus, the largest value P_{ext} can have without a compression taking place is equal to P_{gas}. But an expansion that occurs under these conditions is the reversible expansion. Thus, maximum work is done in the reversible expansion of a gas.

FIRST LAW OF THERMODYNAMICS

The first law of thermodynamics is, in fact, an application of the broad principle known as the Law of Conservation of Energy to the thermodynamic systems. It states that :

the total energy of an isolated system remains constant though it may change from one form to another.

When a system is changed from state A to state B, it undergoes a change in the internal energy from E_A to E_B. Thus, we can write

$$\Delta E = E_B - E_A$$

This energy change is brought about by the evolution or absorption of heat and/or by work being done by the system. Because the total energy of the system must remain constant, we can write the *mathematical statement of the First Law* as:

$$\Delta E = q - w \qquad ...(1)$$

where q = the amount of heat supplied to the system

w = work done by the system

Thus First Law may also be stated as:

the net energy change of a closed system is equal to the heat transferred to the system minus the work done by the system.

To illustrate the mathematical statement of the First Law, let us consider the system 'expanding hot gas'.

The gas expands against an applied constant pressure by volume ΔV. The total mechanical work done is given by the relation

$$w = -P \times \Delta V \quad ...(2)$$

From (1) and (2), we can restate

$$\Delta E = q - P \times \Delta V.$$

We must remember that, the First Law of Thermodynamics, expresses the conversation of energy principle.

HEAT CAPACITIES

The temperature dependence of internal energy and enthalpy depends on the heat capacities at constant volume and constant pressure.

By heat capacity of a system we mean the capacity to absorb heat and store energy. As the system absorbs heat, it goes into the kinetic motion of the atoms and molecules contained in the system. This increased kinetic energy raises the temperature of the system. It q calories is the heat absorbed by mass m and the temperature rises from T_1 to T_2, the heat capacity (c) is given by the expression

$$c = \frac{q}{m \times (T_2 - T_1)} \quad ...(1)$$

Thus, heat capacity of a system is \he heat absorbed by unit mass m raising the temperature by one degree (K or °C) at a specified temperature.

When mass considered is 1 mole, the expression (1) can be written as

$$C = \frac{q}{T_2 - T_1} = \frac{q}{\Delta T} \quad ...(2)$$

where C is denoted as Molar heat capacity.

The molar heat capacity of a system is defined as the amount of heat required to raise the temperature of one mole of the substance (system) by 1 K.

Since the heat capacity (C) varies with temperature; its true value will be given as

$$C = \frac{dq}{dT}$$

where dq is a small quantity of heat absorbed by the system, producing a small temperature rise dT.

Thus the molar heat capacity may be defined as the ratio of the amount of heat absorbed to the rise in temperature.

Units of Heat Capacity

The usual units of the molar heat capacity are calories per degree per mole (cal J K^{-1} mol^{-1}), or joules per degree per mole (J $K^{-1}mol^{-1}$), the latter being the SI unit.

Heat is not a state function, neither is heat capacity. It is, therefore, necessary to specify the process by which the temperature is raised by one degree. The two important types of molar heat capacities are those :

(1) at constant volume; and

(2) at constant pressure.

Molar Heat Capacity at Constant Volume

According to the first law of thermodynamics

$$dq = dE + PdV \qquad ...(i)$$

Dividing both sides by dT, we have

$$\frac{dq}{dT} = \frac{dE + PdV}{dT} \qquad ...(ii)$$

At constant volume dV = 0, the equation reduces to

$$C_V = \left(\frac{dE}{dT}\right)_V$$

Thus the heat capacity at constant volume is defined as the rate of change of internal energy with temperature at constant volume.

Molar Heat Capacity at Constant Pressure

Equation (ii) above may be written as

$$C = \frac{dE}{dT} + \frac{PdV}{dT} \qquad ...(iii)$$

We know $\quad H = E + PV$

Differentiating this equation w.r.t T at constant pressure, we get

$$\left(\frac{dH}{dT}\right)_p = \left(\frac{dE}{dT}\right)_p + P\left(\frac{dV}{dT}\right)_p \qquad \text{...(iv)}$$

comparing it with equation (iii) we have

$$C_p = \left(\frac{dH}{dT}\right)_p$$

Thus, heat capacity at constant pressure is defined as the rate of change of enthalpy with temperature at constant pressure.

Relation Between C_p and C_v

From the definitions, it is clear that two heat capacities are not equal and C' is greater than C_v by a factor which is related to the work done.

Table : Heat Capacities in $JK^{-1}mol^{-1}$ at Constant Pressure.

	a	$b \times 10^3$	$c \times 10^{-5}$
Gases (in Temperature Range 298 to 2000 K)			
He, Ne, Ar, Kr, Xe	20.79	0	0
S	22.01	–0.42	1.51
H_2	27.28	3.26	0.50
O_2	29.96	4.18	–1.67
N_2	28.58	3.76	–0.50
S_2	36.48	0.67	–3.76
CO	28.41	4.10	–0.46
F_2	34.56	2.51	–3.51
Cl_2	37.03	0.67	–2.84
Br_2	37.32	0.50	–1.25
I_2	37.40	0.59	–0.71
CO_2	44.22	8.79	–8.62
H_2O	30.54	10.29	0
H_2S	32.68	12.38	–1.92
NH_3	29.75	25.10	–1.55
CH_4	23.64	47.86	–1.92
TeF_6	148.66	6.78	–29.29

At a constant pressure pan of heat absorbed by the system is used up in increasing the internal energy of the system and the other for doing work by the system. While at constant volume the whole of heat absorbed is utilised in increasing the temperature of the system as there is no work done by the system. Thus increase in temperature of the system would be lesser at constant pressure than at constant volume. Thus C_p *is greater than* C_v.

Liquids (from Melting Point to Boiling Point)			
I_2	80.33	0	0
H_2O	75.48	0	0
NaCl	66.9	0	0
$C_{10}H_8$	79.5	407.5	0
Solids (from 298 K to Melting Point, or 2000 K)			
C (graphite)	16.86	4.77	–8.54
Al	20.67	12.38	0
Cu	22.63	6.28	0
Pb	22.13	11.72	0.96
I_2	40.12	49.79	0
NaCl	45.94	16.32	0
$C_{10}H_8$	–115.9	937	0

Heat capacities are properties of the system. They are directly related to the way the internal energy and enthalpy change with temperature when the volume or pressure of the system is suitable controlled.

THERMAL PART OF THE INTERNAL ENERGY

The thermal part of the internal energy and the enthalpy of an ideal gas can be given a molecular-level explanation.

All the preceding development of internal energy and enthalpy has been "thermodynamic." We could calculate the thermal-energy contribution $U - U_0$ for molecularly simple systems such as ideal gases. Now this thermal energy will be related to the thermodynamic internal energy and enthalpy.

Uo represents the energy of a system when only the lowest available energy levels are occupied. This is the energy that the system would have if the temperature were lowered to absolute zero and the system did not change its physical form. The thermal energy $U - U_0$ is the additional energy that the system would acquire if the temperature were raised from this hypothetical zero-temperature form and the particles distributed themselves throughout the energy levels.

Thermal Enthalpy $H - H_0$

The general relation between enthalpy and internal energy is

$$H = U + PV.$$

For liquids and solids at all ordinary pressures, the change in the PV term is small compared to changes in the H and U terms. As a result, at temperature the enthalpy and internal energy are effectively equal. Thus $H = U$, $H = U_0$, and $H - H_0 = U - U_0$. For standard state thermal enthalpies and internal energies, we have

$$H^o_T - H^o_0 = U^o_T - U^o_0 \qquad \text{(liquid or solid)}$$

For gases, ideal behavior allows the PV term for a sample containing 1 mol of gas molecules to be equated to RT. Then the general enthalpy-internal relation is $H = U + RT$. When only the lowest-energy states are occupied, as occurs at $T = 0$, the RT terms goes to zero, and we have $H_0 = U_0$. It follows, from $H = U + RT$ and $H_0 = U_0$, that $H - H_0 = U - U_0 + RT$ for 1 mol of an ideal gas. The thermal-energy results can be converted to enthalpies by addition of RT. For standard state thermal enthalpies and internal energies we have

$$H^o_T - H^o_0 = (U^o_T - U^o_0) + RT \qquad \text{[ideal gas]}$$

Standard Enthalpy of Formation at Temperatures Other Than 298 K

The enthalpies of formation of many substances have been deduced for the convenient reference temperature of 25°C (298 K). This reference temperature is not the one we would pick if we set out to deal with the molecular basis of energies. Absolute zero, the temperature at which all the molecules would be in their lowest energy states, would be a more natural reference temperature.

We then could obtain the internal energy or enthalpy at other temperatures by simply adding the appropriate thermal energy term. (When we deal with another thermodynamic property, entropy, an even more compelling reason for thinking about the absolute zero and the thermal contributions to a quantity.)

Just as we defined the standard enthalpy of formation $\Delta H^{o}_{f,\,298}$ at 298 K, so we can define the standard enthalpy of formation $\Delta H^{o}_{f,T}$ at any temperature. Some values of $\Delta H^{o}_{f,T}$.

Pay attention to the subscript f on these enthalpies of formation. That signals the fact that the values are for the enthalpies of the compounds compared to the enthalpies of the elements of which they are composed—at the indicated temperature. It follows that the *standard* enthalpies of formation of the elements themselves are zero at any temperature.

Standard Enthalpies of Atomic Species

We need energy data for the free, gaseous atoms to calculate the energy change when the molecules of a substance are broken up into free atoms.

Enthalpy and energy data can be obtained for gaseous atomic substances, These data come, generally, from spectroscopic rather than calorimetric measurements. For diatomic molecules, spectral studies show the energy for the break-up of these molecules into atoms. Results for the original molecules and the atoms produced, all in their lowest-energy, or "ground," states, can be deduced from the spectral data. Thus, we arrive directly at data for $\Delta H^{o}_{f,\,0}$. These energy data for atomic species can be extended to give enthalpy values.

IONIC HYDRATION

This is as far as we can go with a thermodynamic treatment. The fact, however, that the difference between successive values in adjacent columns or rows are approximately constant leads us to recognize that the energy of solution of the ions of an electrolyte, at infinite dilution, can be interpreted in terms of separate contributions made by the ions of the electrolyte.

A variety of evidence suggests that the ions present in an aqueous solution have associated with them water molecules that are said to solvate, or *hydrate*, the ion. The associations involve various numbers of water molecules and occur with varying energies, and the complexes of ion-plus-water molecules persist for varying lengths of time.

Attempts have been made to separate the hydration energy of the ions ot an electrolyte into ionic components. Most start with the calculation of the relative energy required to give an electrical charge to a sphere which represents the ion, in the gas phase compared with the energy required for an ion in the solvent. A result of such a derivation suggests that the hydration energy should be proportional to the square of the charge and inversely proportional to its effective radius in aqueous solution. With such a guide, results like those can be divided into contributions from each of the ions of the electrolyte. One set of values, which can be looked on as based on the assumption of –1090 kJ mol^{-1} for the H^+ ion.

Some of the trends of the data expected from the sizes and charges of the ions. (For more extensive data, a closer look into the arrangement of the outer electrons of the ion and the way these electrons interact with the adjacent water molecules must be taken.) Of special note, however, is the very large value for the proton. Some understanding of this value is provided by an estimate that the energy change for the gas phase reaction

$$H^+(g) + H_2O(g) \rightarrow H_2O + (g)$$

is –760 kJ mol^{-1}. This reaction can be considered to be a step in the solution of a gaseous H + ion. For this step to occur, an energy expenditure of about 40 kJ is needed to form 1 mol of $H_2O(g)$ from liquid water. For the energy change of the entire solution process to be –1090 kJ, the energy change for the solution of 1 mol of H_3O + (g) must be –1090 – (–760 + 40) = –370 kJ. This value is in line with the values for other singly charged ions.

Noteworthy also are the very large solution energy values for the highly charged ions such as Al^{3+}. If these values are divided by 6, a reasonable value for the number of water molecules that can come in direct contact with the ion may lead to hydration energies for each nearest-neighbour water molecule. Thus, for these ions the hydration process must be regarded as involving the formation of bonds with strengths comparable to those of covalent bonds. For the singly charged ions, however, the energy per nearest-neighbour water molecule is considerably less, and a very strong hydrogen bond or ion-dipole association is indicated. Thus, although the ion hydration energies cannot be easily understood, they are clearly significant for the description of ions in aqueous solution.

ENTHALPY CHANGE IN CHEMICAL REACTIONS

Many reactions are not suitable for direct calorimetric study. The internal-energy or enthalpy changes can often be obtained by an indirect method. This indirect procedure was originally suggested by Hess in

1840, and it is often known as *Hess's law of heat summation*. We now recognize that it is merely an application of the first law of thermodynamics.

The indirect determination of the enthalpy change can be illustrated with the reaction in which carbon is converted from graphite to diamond, *i.e.*,

$$C\ (\text{graphite}) \rightarrow C\ (\text{diamond})\ \Delta H = ?$$

Although, this reaction can be made to occur, it is certainly very unsuitable for any direct calorimetric study.

The combustion of both graphite and diamond can be conveniently studied, and these reactions and the enthalpy changes for the combustion are

$$C\ (\text{graphite}) + O_2(g) \rightarrow CO_2(g) \qquad \Delta H^{o}_{298} = -\ 393.51\ \text{kJ}$$

$$C\ (\text{diamond}) + O_2(g) \rightarrow CO_2(g) \qquad \Delta H^{o}_{298} = -\ 395.40\ \text{kJ.}$$

The enthalpy changes of these reactions clearly differ as a result of the different enthalpies of graphite and diamond, as shown by writing

$$-393.51\ \text{kJ} = H_{CO2'} - H_{C(graphite)} - H_{O2}$$

$$-395.40\ \text{kJ} = H_{CO2'} - H_{C(diamond)} - H_{O2}$$

Subtraction of these algebraic equations with cancellation of the enthalpies of CO_2 and O_2 gives, on rearrangement,

$$H_{C(diamond)} - H_{C(graphite)} = -393.51 + 395.40 = +\ 1.89\ \text{kJ.}$$

For the original graphite-to-diamond reaction, we now can write

$$C\ (\text{graphite}) \rightarrow C(\text{diamond}) \qquad \Delta H = +\ 1.89\ \text{kJ.}$$

The same result can be obtained by thinking of the common product, here CO_2, as an intermediate for the reaction in which we are interested. Then equations are written for which the original reactants form this intermediate, and the equation for which this intermediate produces the products of the original reaction.

By this we get the following equations.

$$C\ (\text{graphite}) + O_2(g) \rightarrow CO_2(g) \qquad \Delta H = -\ 393.51\ \text{kJ}$$

and

$$CO_2(g) \rightarrow C\ (\text{diamond}) + O_2(g) \qquad \Delta H = +\ 395.40\ \text{kJ.}$$

In an alternative procedure, both reactions are written to show the formation of the common intermediate, or intermediates. Then the equations are "subtracted " to produce the cancellation that gives the equation for the overall transformation of reactants to products.

Enthalpies, known as standard enthalpies of formation, can be assigned to substances.

Methods have now been outlined for obtaining the enthalpy change that occurs when a chemical transformation takes place. It is not practical to tabulate the enthalpy changes for all *chemical reactions* that have been studied. It is, however, feasible to list enthalpy information for all the *substances* involved in these reactions.

Ion Enthalpies

Our goal is a table of ΔH^o_f values for individual ions. To reach this goal, we must first see that the enthalpy of a dilute solution can, in fact, be treated in terms of contributions attributed to the separate types of ions in the solution. We learn that this can be done for well-behaved dilute solutions by, for example, mixing dilute solutions of HCl and KBr. No energy changes result when the mixture is formed. Such results show that the energy of such dilute solutions can be attributed to separate contributions from the ions present. Each ionic contribution is independent of the other ions present in the solution.

The fact that no ordinary chemical reaction involves only a single type of ion frustrates any attempt to obtain an experimental value for ionic enthalpies. But the basis of this frustration permits us to assign an arbitrary value to the standard enthalpy of any one ionic species.

It is agreed that the value of zero will be assigned to the standard enthalpy of formation of H^+ ions in dilute aqueous solution. That is, we set

$$\Delta H^o_f[H^+(aq)] = 0.$$

Once this step has been taken, the values of the standard enthalpies of other ions can be obtained. For example, the standard enthalpy of the chloride ion at 25°C can be calculated from the data we obtained from the HCl solution process. We have

$$\Delta H^o_f[Cl^-(aq)] = -167.16 - \Delta H^o_f[H^+(aq) = -167.16 - 0 = -167.16 kJ$$

We now have a route to the enthalpies of all other ions.

JOULE THOMSON EXPERIMENT

The phenomenon of producing lowering of temperature when a gas is nude to expand adiabatically from a region of high pressure into a region of low pressure, is known as Joule-Thomson Effect or Joule-Kelvin Effect.

Joule-Thomson Experiment

The apparatus used by Joule and Thomson to measure the temperature change on expansion of a given volume of gas. An insulated tube is fitted with a porous plug in me middle and two frictionless pistons A and B on the sides. Let a volume V_1 of a gas at pressure P_1 be forced through the porous plug by a slow movement of piston A. The gas in the right-hand chamber is allowed to expand to volume V_1 and pressure P_2 by moving the piston B outward. The change in temperature is found by taking readings on the two thermometers.

Most gases were found to undergo cooling on expansion through the porous plug. Hydrogen and helium were exceptions as these gases showed a warming up instead of cooling.

The work done on the gas at the piston A is P_1V_1 and the work done by the gas at the piston B is P_2V_2. Hence, the net work (w) done by the gas is

$$w = P_2P_2 - P_1P_1$$

$$\Delta E = q - w \qquad \text{(First Law)}$$

But the process is adiabatic and, therefore, $q = 0$

$$\therefore \qquad \Delta E = E_2 - E_1 = -w = -(P_2V_2 - P_1V_1)$$

$$\text{or} \qquad E_2 - E_1 = -(P_2V_2 - P_1V_1)$$

Rearranging,

$$E_2 + P_2V_2 = E_1 + P_1V_1$$

$$H_2 = H_1 \quad \text{or } \Delta H = 0.$$

Thus the process in Joule-Thomson experiment takes place at constant enthalpy.

Joule-Thomson Coefficient

The number of degrees temperature change produced per atmosphere drop in pressure under constant enthalpy conditions on passing a gas through the porous plug, is called Joule-Thomson coefficient. It is represented by the symbol μ. Thus

$$\mu = \frac{dT}{dP}$$

If μ is positive, the gas cools on expansion; if μ is negative, the gas warms on expansion. The temperature at which the sign changes is called the inversion temperature. Most gases have positive Joule-Thomson coefficients and hence they cool on expansion at room temperature. Thus liquefaction of gases is accomplished by a succession of Joule-Thomson expansions.

The inversion temperature for H_2 is –80°C. Above the inversion temperature, μ is negative. Thus at room temperature hydrogen warms on expansion. Hydrogen must first be cooled below –80°C (with liquid nitrogen) so that it can be liquefied by further Joule-Thomson expansion. So is the case with helium.

We have shown above that Joule-Thomson expansion of a gas is carried at constant enthalpy. But

$$H = E + PV.$$

Since H remains constant, any increase in PV during the process must be compensated by decrease of E, the internal energy. This leads to a fall in temperature *i.e.,* $T_2 < T_1$. For hydrogen and helium PV decreases with lowering of pressure, resulting in increase of E and T_2, > T_1. Below the inversion temperature, PV increases with lowering of pressure arid cooling is produced.

HEAT AND WORK

When a change in the state of a system occurs, energy is transferred to or from the surroundings. This energy may be transferred as heat or mechanical work.

We shall refer the term 'work' for mechanical work which is defined as force × distance

Units of Work

In CGS system the unit of work is erg which is defined as the work done when a resistance of 1 dyne is moved-through a distance of 1 centimetre. Since the erg is so small, a bigger unit, the *joule* (J) is now used.

$$1 \text{ joule} = 10^7 \text{ ergs}$$

or

$$1 \text{ erg} = 10^7 \text{ J}$$

We often use *kilojoule (kJ)* for large quantities of work

$$1 \text{ kJ} = 1000 \text{ J}$$

The unit of heat, which was used for many years, is *calorie* (cal). A calorie is defined as the quantity of heat required to raise the temperature of one gram of water by 1°C in the vicinity of 15°C.

Since heat and work are interrelated, *SI unit of heat is the joule* (J).

$$1 \text{ joule} = 0.2390 \text{ calories}$$

$$1 \text{ calorie} = 4.184 \text{ J}$$

$$1 \text{ kcal} = 4.184 \text{ kJ.}$$

The symbol of heat is q. If the heat flows from the surroundings into the system to raise the energy of the system, it is taken to be positive, +q. If heat flows from the system into the surroundings, lowering the energy of the system, it is taken to be negative, –q.

The symbol of work is w. If work is done on a system by the surroundings and the energy of the system is thus increased, it is taken to be positive, +w. If work is done by the system on the surroundings and energy of the system is decreased, it is taken to be negative, –w.

Summary of Sign Conventions

Heat flows into the system, q is +ve	Heat flows out of the system, q is –ve
Work is done on the system, w is +ve	Work is done by the system, w is –ve.

In physics, mechanical work is defined as force multiplied by the distance through which the force acts. In elementary thermodynamics the only type of work generally considered is the work done in expansion (or compression) of a gas. This is known as *pressure-volume work* or *PV work* or *expansion work.*

SECOND LAW OF THERMODYNAMICS

The second law of thermodynamics states that:

Whenever a spontaneous process takes place, it is accompanied by an increase in the total energy of the universe.

More specifically, we take the term 'universe' to mean the system and the surroundings. Thus,

$$\Delta S_{univ} = S_{syst} + S_{surr}$$

The second law, as stated above, tells us that when an irreversible spontaneous process occurs, the entropy of the system and the surroundings increases. In other words $\Delta S_{univ} = 0$. When a reversible process occurs, the entropy of the system remains constant $\Delta S_{univ} = 0$. Since the entire universe is undergoing spontaneous change, the second law can be most generally and concisely stated as :

The entropy of the system is constantly increasing.

Although, the second law can be stated in a number of ways, all statements can be shown to generalize our knowledge about the direction in which natural processes occur. The second law sums up our experiences with the direction of spontaneous processes, just as the first law summed up our experience with energy. The general statements of the second law, as the conservation-of-energy statement of the first law, are not immediately applicable to chemical problems. After general statements of the law are presented, we show that the law can be expressed in a chemically useful form.

Two important statements of the second law have been given. One, due to Lord Kelvin, is that *it is impossible by a cyclic process to take heat from a reservoir and convert it into work without at the same time transferring heat from a hot to a cold reservoir*. This statement can be illustrated by the fact that a ship cannot derive mechanical energy from the energy in the sea on which it sails. With a moment's thought about all types of engines, you will see that there is always both a hot source and a cold sink. A steam engine, for example, could not operate if it were not for the high pressure and high temperature of the steam *compared* with those of the surroundings.

The Kelvin statement of the second law is related to equilibria when it is realized that mechanical energy can be obtained from a system only when the system is not already at equilibrium. If a system is at equilibrium, no process tends to occur spontaneously and there is nothing to harness. A nonchemical example is the production of hydroelectric power. Here mechanical energy is obtained when the spontaneous tendency of water to flow from a high to a low level is harnessed. Lord Kelvin's statement recognizes that the spontaneous process is the transfer of thermal energy from a higher to a lower temperature and that only from such a spontaneous process can mechanical energy be obtained.

The second classic statement, given by Clausius, is that *it is impossible to transfer heat from a cold to a hot reservoir without at the same time converting a certain amount of work to heat*. This statement is illustrated by the operation of a refrigerator. Again we recognize that the spontaneous transfer of thermal energy is from a high to a low temperature and that the reverse is possible only when mechanical energy is expended.

Neither the Kelvin nor the Clausius statement follows from the conservation-of-energy principle of the first law. No conflict with the first law would occur if thermal energy were completely converted to mechanical energy or if thermal energy were to be simply transferred from a lower to a higher temperature. The second law is a different generalization about energy from that provided by the first law.

ENTROPY

Entropy is a thermodynamics state quantity that is the measure of the randomness or disorder of the molecules of the system.

For many years scientists believed that only exothermic changes resulting in a lowering of internal energy or enthalpy could occur spontaneously. But melting of ice is an endothermic process and yet occurs spontaneously. On a warm day, ice melts by itself. The evaporation of water is another example of a spontaneous endothermic process. Thus arose the need of inventing another driving force that affects the spontaneity. This was known as the entropy change, ΔS.

The symbol of entropy is S, while the change in disorder accompanying a process from start to completion is represented by ΔS. The entropy of a system is a state function and depends only on the initial and final states of the system. The change in entropy, ΔS, for any process is given by the

$$\Delta S = S_{final} - S_{initial}$$

when $S_{final} > S_{initial}$, ΔS is positive.

Entropy Changes of the Mechanical Surroundings

We are at liberty to ascribe any self-consistent features as we work toward the entropy property. To this end we specify that for all processes

$$\Delta S_{mech} = 0$$

or

$$dS_{mech} = 0.$$

(The basis for this can also be seen by anticipating the interpretation of entropy as a measure of the molecular-level disorder. Changes in the energy of the mechanical surroundings, a weight-and-pulley system, for example, produce no change in the molecular-level disorder.)

Entropy Change of the System

The following statement lets us deduce the entropy change in the system: *When a process is carried out reversibly, the entropy change in the universe of the system is zero.*

With this statement we can use the entropy changes in the surroundings to determine the entropy change in the system. (Recall that we also used a conservation statement to deduce the energy change in the system from the energy changes in the surroundings.) Now three examples will show you how to calculate the change in the entropy of the system. (Notice that you can do these calculations even though at this stage you do not " know what entropy is" nor do you see what value there is in dealing with this property.) In these examples the entropy change is calculated for a phase change, and a volume or pressure change.

Entropy Change for a Change in Volume or Pressure of an Ideal Gas

Determine the entropy change for the isothermal expansion of n mol of an ideal gas at temperature T from a volume (V_1 to a volume V_2).

Consider the reversible isothermal expansion of an ideal gas. Since the temperature is constant, the internal energy does not change. The thermal surroundings must provide an amount of energy equal to that gained by the mechanical surroundings. We can calculate the change in the energy of the mechanical surroundings as

$$\Delta U_{mech} = \int P dV = \int_{V_1}^{V_2} \frac{nRT}{V} dV = nRT \ln \frac{V_2}{V_1}$$

Now, with $\Delta U = 0$ and therefore $\Delta U_{therm} = \Delta U_{mech}$,

$$\Delta U_{therm} = -\ nRT \ln \frac{V_2}{V_1}$$

This leads us to

$$\Delta S_{therm} = \frac{\Delta U_{therm}}{T} = -\ nR \ln \frac{V_2}{V_1}$$

For the entropy change of the universe of this reversible process to be zero, that is, for $\Delta S + \Delta S_{therm} = 0$, $\Delta S = -\Delta S_{therm}$. Thus

$$\Delta S = +nR \ln \frac{V_2}{V_1}$$

An expanded gas sample has more entropy than did the original sample. Once again we can summarize the considerations that led to this conclusion by a convenient expression. We have

$$\Delta S = nR \ln \frac{V_2}{V_1} = nR \ln \frac{P_1}{P_2} \qquad \text{(n mol of ideal gas)}$$

System Entropy is a Property of the System

One important justification of the arbitrary steps we have taken must be made. Suppose the liquid water at 100°C was converted to water vapour at the normal boiling point of 100°C by a different path for example, imagine reversible processes in which the water was cooled to 25°C and converted to an equilibrium vapour at that temperature, then the vapour was heated to 100°C, and finally the hot vapour was compressed to 1 atm. (From the above examples you can see that you could calculate the entropy change for each step.) Would we necessarily have the same net entropy change as we calculated for the direct conversion?

This specific question can be answered by doing the entropy calculations for all the steps of the indirect path. The components of the total entropy change by two different paths show that 1 mol of water vapour at 100°C and 1 atm has about 132 J K^{-1} more entropy than does 1 mol of liquid water at 25°C and 1 atm.

Table Standard entropies of some substances (25°C, 1 atm).

	Entropy, S°			Entropy, S°	
Substance	**cal mol^{-1} K^{-1}**	**$Jmol^{-1}K^{-1}$**	**Substance**	**cal $mol^{-1}K^{-1}$**	**$Jmol^{-1}K^{-1}$**
Ag (s)	41.32	172.9	H_2 (g)	31.21	130.6
AgCl (s)	58.5	24.5	H_2O (g)	45.11	188.7
Al (s)	6.77	28.3	H_2O (*l*)	16.72	69.96
Al_2O_3 (s)	12.19	51.0	HCl (g)	44.62	186.7
C (s, graphite)	0.58	2.4	HNO_3 (*l*)	37.19	155.6
CO (g)	47.30	197.9	H_2SO_4 (*l*)	37.5	157.0
CO_2(g)	51.06	213.6	Hg (*l*)	18.2	76.1
CH_4 (g)	44.50	186.2	K (s)	38.30	160.2
CH_3OH (*l*)	30.3	126.8	KCl (s)	57.24	239.5
$CO(NH_2)_2$ (s)	25.0	104.6	K_2SO_4 (s)	42.0	176.0
C_2H_2 (g)	48.0	200.8	N_2 (g)	45.77	191.5
C_2H_6 (g)	54.85	229.5	NH_3 (g)	46.01	192.5
Ca (s)	36.99	154.8	Na (s)	36.72	153.6
$CaCO_3$ (s)	22.2	92.9	NaCl (s)	17.30	72.88
Cl_2 (g)	53.29	223.0	O_2 (g)	49.0	205.0
Fe (s)	6.5	27.0	S (s)	7.62	31.9
Fe_2O_3 (s)	21.5	90.0	SO_2 (g)	59.40	248.5

This provides a specific illustration that the procedure we have used for calculating entropy changes leads us to changes in a property of the system. (That is not a necessary consequence of arbitrary procedures. If we had introduced entropy by beginning with $\Delta U_{therm}/T^2$ instead of $\Delta U_{therm}/T$, we would have found that the "entropy" we calculated for the system would be different for each different process we considered.)

There is a general proof that our procedure for calculating entropy differences is independent of the "path" and, therefore, that entropy is a property of the system. This proof is developed next.

THE CARNOT CYCLE

In 1824 Sadi Carnot proposed a theoretical heat engine to show that the efficiency was based upon the temperatures between which it operated. Carnot's imaginary engine could perform a series of operations between temperatures T_1 and T_2, so that at the end of these operations the system was restored to the original state. *This cycle of processes which occurred under reversible conditions is referred to as the Carnot cycle.* The medium employed in operating Carnot's engine was one mole of an ideal gas which could be imagined to be contained in a cylinder fitted with a frictionless piston.

The Carnot cycle comprises four operations or processes.

(1) Isothermal reversible expansion

(2) Adiabatic reversible expansion

(3) Isothermal reversible compression

(4) Adiabatic reversible compression.

The above four processes are shown in the indicator diagram of Carnot cycle.

SOME SPONTANEOUS PROCESSES

A process, which proceeds of its own accord, without any outside assistance, is termed, a spontaneous or natural process. *The reverse process which does not proceed on its own, is referred to as a nonspontaneous or unnatural process.*

In general, *the tendency of a process to occur naturally is called the spontaneity.*

Rolling Ball : A ball rolls down-hill spontaneously but it will not roll uphill unless work is done on it.

Heat Flow : When two balls of metal, one hot and one cold, are connected, heat flows spontaneously from the hot ball to the cold one, never from cold to hot. It requires work to transfer heat from one place to the other, say, by means of refrigerator pump.

Gas Flow : When a vessel containing a gas is connected to another evacuated vessel, the gas spreads throughout spontaneously unless the pressure is the same in both the vessels. The reverse process of compressing the gas into the original vessel cannot occur unless work is done on it.

Gas Mixing : When two gases are mixed, they spontaneously form a gaseous mixture. Gas mixtures do not sort themselves out so that pure gases concentrate at opposite ends of the container.

Criteria of Spontaneity

Some important criteria of spontaneous physical and chemical changes are listed below :

A spontaneous change is one-way or unidirectional. For reverse change to occur, work has to be done.

For a spontaneous change to occur, time is no factor. A spontaneous reaction may take place rapidly or very slowly.

If the system is not in equilibrium state (unstable), a spontaneous change is inevitable. The change will continue till the system attains the state of equilibrium.

Once a system is in equilibrium state, it does not undergo any further spontaneous change in state if left undisturbed. To take the system away from equilibrium, some external work must be done on the system.

THIRD LAW OF THERMODYNAMICS

That the entropies of substances at 0 K can be assigned the value of zero is the third law of thermodynamics.

All substances have the same entropy at absolute zero. (We will come, shortly, to a more careful statement to this effect.) This generalization is related to another, quite different, observation.

A variety of chemical and physical phenomena can best be studied at very low temperatures. As a result, techniques have been developed to produce these temperatures. Liquid nitrogen, which boils at 77 K at 1-bar pressure, is now a common industrial substance. Liquid nitrogen can be used to cool helium which can then be compressed, cooled, and expanded to yield liquid helium. The normal boiling point of liquid helium is 4 K. Temperatures somewhat below 1 K can be obtained by boiling off some of the helium at reduced pressures. Still lower temperatures can be reached by magnetization- demagnetization procedures, with liquid helium drawing off the released thermal energy. In this way temperatures down to about 10^{-4} K have been reached.

From such low-temperature studies comes the realization that any method that can be used to lower the temperature "peters out" as the temperature approaches absolute zero. A summation of the experiences of those who attempt to reach lower and lower temperatures is that *the absolute zero of temperature is unattainable.*

This generalization is related to the indication that the entropy changes for all reactions would be zero if the reactions occurred at absolute zero. Suppose product substances C and D had, together, a greater entropy than reactant substances A and B. Suppose, this entropy difference persisted through low temperature and on down to absolute zero. The $A + B \rightarrow C + D$ reaction could then be imagined to occur reversibly at some very low temperature and in doing so to reduce the entropy, and the energy, of the thermal surroundings. If substances had different entropies at absolute zero, temperatures could be reduced to, and *below*, absolute zero.

Thus, there is an equivalence between the idea that absolute zero cannot be reached and the generalization that the entropy change for all reactions would be zero at absolute zero. This conclusion must be restricted to materials that are in the thermodynamically most stable state for this temperature range. (One finds, for example, that many materials are frozen into a metastable glassy state as the temperature is reduced, and this state may persist even as absolute zero is approached because of the slowness with which the crystalline form is produced. The entropy of the glassy state could be different, in fact higher, than that of the crystal at absolute zero. Since the metastable state cannot be converted directly to the stable state by a reversible process, this entropy difference could not be used in attempts to reach absolute zero.)

We come to the chemically useful statement of the third law of thermodynamics, quoted from the classic thermodynamics text by Lewis and Randall: "*If the entropy of each element in some crystalline state be taken as zero at the absolute zero of temperature, every substance has a finite positive entropy; but at the absolute zero of temperature the entropy may become zero, and does so become in the case of perfect crystalline substances.*"

The third law makes it possible to assign entropy values, described as *absolute entropies*, to chemical compounds.

FUGACITY AND ACTIVITY

Consider a system composed of liquid water and its vapour. Liquid water has a tendency to escape into the vapour phase while the vapour tends to escape the gaseous state and come into the liquid phase by condensation. When the system is in equilibrium, these two escaping tendencies become equal and we observe a constant vapour pressure at a constant temperature. In general, it may be stated that each substance in a given state has a tendency to escape from that state and this escaping tendency denoted by f is called fugacity. It is related to the free energy content (G) by the expression

$$G = RT\ln f + B$$

where B is a constant depending upon the temperature and the nature of the substance. It is not possible to evaluate B since the absolute values of the free energy are not known. To circumvent this difficulty, all free energy measurements for any given substance are referred to as standard reference point. If we represent by G^o the free energy per mole and f^o the fugacity in this standard state, then G^o is given by

$$G^o = RT \ln f + B.$$

If G is the free energy of the substance in any state, then the free energy difference between this state and the standard state is given by

$$G - G^o = RT \ln f^o + B$$

or

$$G = G^o + RT \ln f/f^o \qquad ...(i)$$

The ratio f/f^o is called *activity* and is denoted by the symbol a. The activity of any substance may therefore, be defined as the *ratio of fugacity of the substance in the given state to the fugacity of the same substance in the standard state.*

$$G = G^o + RT \ln a$$

In the standard state, $G = G^o$

$\therefore$ $RT \ln a = 0 \quad \text{or } a = 1$

i.e., in the standard state the activity of a substance is equal to unity. In any other state the value of activity will depend upon the difference $(G - G^o)$. The difference in free energy per mole caused on passing from one state in which the free energy is G_1 and the activity a_1 to another state in which these are G_2 and a_2 respectively, is given by the expression

$$\Delta G = G_2 - G_1 = (G^o + RT\ln a_2) - (G^o + RT\ln a_1)$$

or
$$\Delta G = RT\ln\frac{a_2}{a_1}$$

The similarity between the above equation and equation

$$\Delta G = RT\ln\frac{P_2}{P_1}$$

suggests that *activity is the thermodynamic counterpart of the gas pressure.*

For the standard state of any gas at the given temperature, the fugacity is taken as equal to unity, viz, $f^o = 1$ and on the basis of this definition the activity of any gas becomes equal to fugacity

$$a = \frac{f}{f^o} = \frac{f}{1} = f.$$

The equation (i) can, therefore, be written as

$$G = G^o + RT\log_e f.$$

For an *ideal gas* the fugacity is equal to pressure and f/P = 1. For a real gas, the fugacity is not equal to P and the ratio f/P varies. It is observed, however, that on decreasing the pressure, the behaviour of the gas approaches that of an ideal gas. It may be stated, therefore, that f approaches P as P approaches zero

i.e.,
$$\lim_{P\to 0}\frac{f}{P} = 1$$

or
$$\frac{f}{P} = 1 \text{ as } P \to 0.$$

The ratio f/P is called *activity coefficient* of a gas and is represented by the symbol y. It gives a direct measure of the extent to which any gas deviates from ideal behaviour at any given pressure and temperature for the farther this ratio is from unity, the greater is the non-ideality of the gas.

WORK AND FREE ENERGY FUNCTIONS

Besides heat content (H), internal energy (E) and entropy (S), there are two other thermodynamic functions depending upon the state of the system which utilize E, H or S in their derivation and are more convenient for use. These are *Work* and *Free energy functions* represented by A and G respectively.

The Work Function

The work function (A) is defined by

$$A = E - TS$$

where E is the energy content of the system, T is its absolute temperature and S its entropy. Since E, F and S depend upon the thermodynamic state of the system only and not on its previous history, it is evident that the function A is also a single valued function of the state of the system.

Consider an isothermal change at temperature T from the initial state indicated by subscript 1 to the final state indicated by subscript 2, so that

$$A_1 = E_1 - TS_1 \quad ...(1)$$

and

$$A_2 = E_2 - TS_2 \quad ...(2)$$

Subtracting (1) from (2), we have :

$$A - A_1 = (E_2 \, E_1) - T(S_2 - S_1)$$

$$\Delta A = \Delta E - T\,\Delta S \quad ...(3)$$

where ΔA is the increase in function A, ΔE is the corresponding increase in internal energy and ΔS is the increase in the entropy of the system.

Since $\Delta S = q_{rev}/T$ where q_{rev} is the heat taken up when the change is carried out in a reversible manner at a constant temperature, we have:

$$\Delta A = \Delta E - q_{rev} \quad ...(4)$$

According to the first law of thermodynamics, $\Delta E = (q_{rev} - w_{rev})$

or

$$w_{rev} = \Delta E - q_{rev} \quad ...(5)$$

Substituting this value in equation (4), we get

$$-\Delta A = w_{rev}$$

i.e., decrease in the work function A in any process at constant temperature gives the maximum work that can be obtained from the system during any change.

6

Raman Spectroscopy

INTRODUCTION

Rotational-energy changes can also be studied by means of a scattering technique known as *Raman spectroscopy*. The sample is irradiated with a monochromatic beam of, usually, visible radiation. Laser beams are especially suitable. The wavelength of the beam is chosen so that the beam is not strongly absorbed by the sample. The beam does produce an induced dipole in the molecule, and this allows transfer of energy between the molecules of the sample and the beam.

The classical picture again suggests some features of the interaction between the radiation and the rotating molecule. The electric field $E_0 \cos 2\pi vt$ of the radiation passing a molecule will distort the electronic structure and produce an induced dipole in the direction of the electric field. If the polarizability α is introduced as the proportionality constant between the electric field and the induced dipole moment, we can write

$$\mu_{ind} = \alpha E_0 \cos 2\pi vt.$$

For a rotating molecule that is not spherically Symmetric, the polarizability along a direction will vary about an average value with a frequency equal to twice that of the rotational frequency. We can express this as

$$\alpha = \alpha_{av} + \Delta\alpha \cos 2(2\pi v_R)t$$

The induced dipole varies with time according to

$$\mu_{ind} = [\alpha_{av} + \Delta\alpha \cos 2(2\pi v_R)t](E_0 \cos 2\pi vt)$$

The trigonometric relation $2 \cos \theta \cos \phi = \cos(\theta + \phi) + \cos(\theta - \phi)$ shows that this expression is equivalent to

$$\mu_{ind} = \alpha_{av} E_0 \cos 2\pi vt + (\Delta\alpha)E_0 [\cos 2\pi(v + 2v_R)t + \cos 2(2\pi v_R)t]$$

SPECTRA OF CONJUGATED SYSTEMS

For larger and generally shaped molecules, detailed descriptions of the states involved in an electronic transition often cannot be achieved. Spectroscopic techniques still provide a considerable insight into electronic properties.

The electronic absorptions of organic compounds, usually found in the ultraviolet region, can often be identified with a group within the molecule. Electrons in single covalent bonds such as C—C and C—H require very large energies to produce electronic excitation. Saturated hydrocarbons absorb only very high energy radiation, usually beyond 160 nm, far in the ultraviolet region. A simple olefin, however, has an absorption band at around 170 nm, and this can be attributed to the excitation of the n electrons from the electron-paired bonding configuration to a high-energy, or antibonding, state. Such a transition is referred to as a $\pi \rightarrow \pi^*$ *transition*, the asterisk implying an antibonding orbital.

Some molecules have electronic configurations which can be altered in different ways to lead to an excited, or high-energy, electronic state. This situation arises, for example, with compounds containing a carbonyl group C—Ö:. For such a group the possibility of exciting the π electrons to the excited π^* state exists, as with an olefin, to give a $\pi \rightarrow \pi^*$ transition. Alternatively, the non-bonding electrons of the oxygen might be excited to the higher-energy π^* electron state, and the absorption would then be characterized as an $n \rightarrow \pi^*$ transition, where the n signifies a nonbonding electron.

Not all electronic transitions of organic compounds occur in the ultraviolet region. The occurrence of coloured compounds indicates absorption of radiation in the visible spectrum. Such absorption requires the electronic energy levels to be more closely spaced than in most molecules. The most common type of organic molecule that absorbs in the visible region, *i.e.,* is coloured, consists of a conjugated system, frequently involving aromatic rings. The qualitative explanation for the closer spacing that results from the delocalization of the conjugated electrons is most easily given by regarding such electrons as being free particles within the potential box of the molecule. For sufficiently long " boxes," the electronic energy spacing is small enough to bring the absorption of radiation into the visible part of the spectrum. If resonance structures are drawn, each carbon-carbon bond along the chain has appreciable double-bond character. The π electrons are therefore not localized but are relatively free to move throughout the entire carbon skeleton. This suggests that the skeleton be considered as a roughly uniform region of low potential bounded at the ends of the molecule by regions of infinitely high potential. The resulting square potential well is to be the receptacle of the 22 π electron. The energies are given by

$$\varepsilon = \frac{n^2h^2}{8ma^2} \qquad n = 1, 2, 3, \ldots$$

where a is the effective length of the molecule. Two electrons, one with each spin direction, can be placed in each square-well orbital. The electron energies described by this molecular-orbital approach can be explained.

The chief merit of this treatment of conjugated systems is that it offers an easy way to relate the wavelength of light absorbed to the length of the conjugated system. The visible-region absorption of carotene consists of a band centered at 451 nm. Absorption at this wavelength implies the absorption of quanta of energy 4.41×10^{-19} J. According to the free-electron model for the π-electron system, this result can be related to the energy difference

$$\varepsilon_{12} - \varepsilon_{11}(12^2 - 11^2)\,\frac{h^2}{8ma^2}$$

Substitution of numerical values leads to a value of a, the length of the region in which the electrons are free to move, of 1.77 nm or 1770pm. This value, as we shall see when experimental methods for determining molecular dimensions are developed in the following chapters, is very close to that expected for the extended conjugated carbon chain of the β-carotene molecule.

NUCLEAR MAGNETIC-RESONANCE SPECTROSCOPY

The frequency of the radiation that corresponds to the nuclear magnetic-energy-level spacings and the weakness of the radiation absorption that must be expected lead to a spectrometer of a radically different kind from those prism instruments used for electronic and vibrational spectral analyses. The principal magnetic field acts on the nuclei of the sample to produce energy levels. Transitions between these levels are stimulated by radiation from the radio-frequency transmitter, which sends out electromagnetic radiation from the transmitter coil. Radiation will be absorbed and emitted by the sample if the frequency of the radiation is such that the quanta of radiation have an energy matching the nuclear energy-level spacing. The receiver coil, which is oriented at right angles to the transmitter, receives no signal unless the sample provides this coupling with the transmitter. The signal from the receiver coil can be displayed on an oscilloscope or a recorder. This indication of the operation of an nmr spectrometer implies that a fixed magnetic field is imposed on the sample and that the frequency of the radiation is varied.

The nuclear energy-level splitting depends on the nuclear magnetic moment and the magnetic field strength. The experimental results indicate that even if the absorption of only hydrogen atoms is studied, a number of closely spaced absorptions are observed.

ELECTRON SPIN SPECTROSCOPY

Unpaired electrons can be given spin states with different energies by the application of the magnetic field. These states are studied in electron-spin resonance spectroscopy.

The electron like the proton has a half unit of spin angular, momentum. The magnetic moment off the election due to its spin angular momentum is expressed as $(g_e I)\mu_B$. The electron "g factor" relates, the magnetic moment of the electron to its spin quantum number. The value of 'g' for a free electron in most atomic and molecular species is between 1.9 and 2.1. The spin quantum number I of the electron is ½.

The μB term is *Bohr magneton* $\mu_B = e\hbar/2m_e$.), to have a value of $9.2741 \times 10^{-24} Cs^{-1}m^2$, or JT^{-1}.

Notice that the Bohr magneton is about 1000 times greater than the nuclear magneton, so the spin magnetic moment of the electron is about 1000 times greater than nuclear magnetic moments. The energy of interaction of the magnetic moment of the unpaired electron with the applied field will be greater than the corresponding interaction between the nuclear magnetic moment and the applied field.

The energy separation between electron spin states in the relatively low magnetic field of 0.30 T (3000 G) is calculated, by analogy with equation as

$$\begin{aligned} h\nu &= g_e \mu_B B \\ &= 2(9.2741 \times 10^{-24}\ J\ T^{-1})\ (0.30T) \\ &= 5.6 \times 10^{-24}\ J. \end{aligned}$$

Radiation with a frequency $(5.56 \times 10^{-24}\ J)/(6.6262 \times 10^{-34}\ Js) = 0.84 \times 10^{10} s^{-1}$, and a wavelength of 35 mm, can stimulate transitions between the spin states for electrons in this magnetic field. Such radiation occurs in the microwave region.

The most prominent and revealing feature of esr spectra is the splitting caused in the transition between the two electron-orientation states by the interaction of the magnetic moment of the spinning electron with the magnetic moments of those nuclei in the molecule which have magnetic moments. The electron-spin energies and the splitting of these energies due to the nitrogen nucleus, which has one unit of spin. Transitions occur which change the orientation of the electron spin relative to the applied magnetic field. The interaction between the electron and the magnetic nucleus is sufficiently small that the transitions do not also change this magnetic-moment direction. The observed spectrum does in fact show three absorption bands.

Because of the experimental arrangement used in esr, the derivative of the usual spectral absorption or emission curve. The splitting due to interactions with the magnetic nuclei can be treated in much the same way as the nuclear magnetic interactions in nmr spectroscopy. The odd electron can move throughout the molecule and it experiences the effect of the nuclear moments of the four equivalent hydrogens. The esr spectroscopy provides a powerful tool tor the study of chemical species with unpaired electrons. It gives information not only about the presence and number of such electrons, as measurements of paramagnetism often do, but also about the distribution of the electron in the molecule. The splitting due to interactions with the nuclear magnetic moments of atoms of molecules or ions depends on the distribution of the odd electron throughout the molecule. Details of electronic configuration in free-radical-type molecules are one of the important features treated by esr.

PHASE RULE

A phase may be defined as: *any homogeneous part of a system having all physical and chemical properties the same throughout.*

A system may consist of one phase or more than one phases.

(1) A system containing only liquid water is one-phase or 1-phase system (P = 1).

(2) A system containing liquid water and water vapour (a gas) is a two-phase or 2-phase system (P = 2).

(3) A system containing liquid water, water vapour and solid ice is a three-phase or 3-phase system.

A system consisting of one phase only is called a *homogeneous system*. A system consisting of two or more phases is called a *heterogeneous system*. Ordinarily three states of matter gas, liquid, and solid are known as phases. However in *phase rule*, a uniform part of a system in equilibrium is termed a 'phase'. Thus a liquid or a solid mixture could have two or more phases.

Let us consider a few examples to understand the meaning of the term phase as encountered in phase rule.

Pure Substances : A pure substance (solid, liquid, or gas) made of one chemical species only, is considered as one phase. Thus, oxygen (O_2), benzene (C_6H_6), and ice (H_2O) are all 1-phase systems. *It must be remembered that a phase may or may not be continuous*. Thus, whether ice is present in one block or many pieces, it is considered one phase.

Mixtures of Gases : All gases mix freely to form homogeneous mixtures. Therefore any mixture of gases, say O_2 and N_2, is a 1-phase system.

Miscible Liquids : Two completely miscible liquids yield a uniform solution. Thus a solution of ethanol and water is a 1-phase system.

Non-Miscible Liquids : A mixture of two non-miscible liquids on standing forms two separate layers. Hence a mixture of chloroform ($CHCl_3$) and water constitutes a 2-phase system.

Aqueous Solutions : An aqueous solution of a solid substance such as sodium chloride (or sugar) is uniform throughout. Therefore, it is a 1-phase system.

However, a saturated solution of sodium chloride in contact with excess solid sodium chloride is a 2-phase system.

Mixtures of Solids :

(i) By definition, a phase must have throughout the same physical and chemical properties. Ordinary sulphur as it occurs in nature is a mixture of monoclinic and rhombic sulphur. These allotropes of sulphur consist of the same chemical species but differ in physical properties. Thus, a mixture of two allotropes is a 2-phase system.

(ii) A mixture of two or more chemical substances contains as many phases. Each of these substances having different physical and chemical properties makes a separate phase. Thus, a mixture of calcium carbonate ($CaCO_3$) and calcium oxide (CaO) constitutes two phases.

Let us consider the equilibrium system; the Decomposition of Calcium carbonate. When calcium carbonate is heated in a closed vessel, we have

$$\underset{\text{(solid)}}{CaCO_3} \rightleftharpoons \underset{\text{(solid)}}{CaO} + \underset{\text{(gas)}}{CO_2}$$

There are two solid phases and one gas phase. Hence it is a 3-phase system.

PRESSURE TEMPERATURE PHASE DIAGRAMS

The phases present in a one-component system at various pressures and temperatures can conveniently be studied with the help of following diagram. In the diagram, the line labeled TC shows the pressures and temperatures at which liquid and vapor exist in equilibrium. It is a vapor-pressure curve. At temperatures higher than that of point C, the critical-point, liquid-vapor equilibria do not occur. Thus this liquid-vapor equilibrium line terminates at C.

Consider the changes that occur as a pressure or temperature change results in the system moving across the line TC. From point 1, for example, the temperature can be lowered to get to point 2, or the pressure can be raised to get to point 3. In either process one crosses the liquid-vapor equilibrium line in the direction of condensation from vapor to liquid. Notice, however, that if a sample is carried from point 1 to point 2 or point 3 by a path that goes around C, no phase change will occur.

Line TB gives the temperatures and pressures at which solid and vapor are in equilibrium; *i.e.,* it is the curve for the vapor pressure of the solid. Line TA gives the temperatures and pressures at which ice and liquid are in equilibrium; *i.e.,* it shows the melting point of ice as a function of pressure. Liquid water can be cooled below its freezing point to give, as indicated by the dashed line TD, *supercooled* water. Supercooled water represents a *metastable* system. It owes its existence to the fact that the rate of formation of ice has been interfered with by the use of a very clean sample of water and a smooth container.

The phase behavior of water at very high pressures. Many new solid phases, corresponding to ice with different crystal structures, are encountered. The occurrence of different crystalline forms of a given compound is fairly common and is known as polymorphism. It is particularly remarkable that the melting point of ice VII, which exists above about 20,000-bar pressure, is over 100°C.

The occurrence of a single phase corresponds to an area on a P-versus-T diagram; *i.e.*, both these variables can be arbitrarily assigned, within limits, without the appearance of a second phase. When two phases are in equilibrium, the diagram shows a line indicating that either P or T may be fixed but that when one is fixed, the other must be such that the system is somewhere on the phase-equilibrium, line. Finally, three phases can exist together, and this occurs at a point on the diagram. Nowhere on the diagram do four phases coexist.

The most important three-phase equilibrium point, called the *triple point*, is that shown by ordinary ice, liquid, and vapor. The temperature and pressure at the triple point are completely determined by the system itself. The most familiar material, water, that we have used as an illustration of P-versus-T phase diagrams is, in some ways, not at all representative. More suitable, in this regard, is one in which the solid-liquid equilibrium line, TA has a positive slop phase whether it is in a single block or subdivided into fine chips, but this subdivision must not be carried to molecular dimensions. A solution in which there are two chemical species, for example, is to be considered as one phase, even though subdivision to a molecular scale would reveal that it was not "uniform throughout."

PHASE EQUILIBRIA

Phase equilibria can be described interms of the number of phases, components and degrees of freedom. A phase may consist of any amount, large or small of material and may be in one unit or subdivided into a number of smaller units. Thus ice represents a phase.

Of particular importance is the *number of phases* P present in a system. Because of the complete mutual solubility of gases, only one gaseous phase can exist in any system. Some liquids are insoluble in each other, and a number of different liquid phases may therefore exist in a system at equilibrium. Different solids, whether they have different chemical compositions or the same chemical composition but different crystal structures, constitute different phases. It is necessary now to consider what information must be given to specify the chemical composition of a system. In this connection, the familiar word components is used, but a strict definition is attached to it. The *number of components* C is the least number of independently variable chemical species necessary to describe the composition of each and every phase of the system.

The composition of a solution of sugar in water, for example, is described by specifying that sugar and water are present. There are two components. If such a solution is cooled, a pure solid sugar phase may begin to separate out.

According to the definition, the system still has two components even if the solid phase contains only one chemical species. Some special care is required when the system involves species in chemical equilibrium with each other. The number of species that can be arbitrarily varied in a solution of acetic acid in water is two, A number of equilibria are set up in such a system; in particular,

$$CH_3COOH + H_2O \rightleftharpoons CH_3COO^- + H_3O^+$$

An overstrict attention to the possible equilibria among the species of a system should be avoided. Consider, for example, the gaseous system of water vapor, hydrogen, and oxygen. In the presence of an electric arc or suitable catalyst, the equilibrium

$$2H_2O \rightleftharpoons 2H_2 + O_2$$

is readily established. Under such conditions the system has two components since specification of two species implies that at equilibrium the third will be present. Alternatively, one can say that the concentration of any two species could be arbitrarily set, but that the concentration of the third would then be fixed and could in fact be calculated from the equilibrium constant of the reaction. At room temperature and in the absence of a catalyst, however, this equilibrium is established so slowly that for all practical purposes the reaction connecting the three species can be ignored.

DETERMINATION OF TRANSITION POINT

The temperature at which a polymorphic substance changes from one form to another is known as the transition temperature or transition point. The transition point can be determined by measuring a change in physical properties, such as colour, density solubility etc. Now let us study one after one.

Colour Change

If a little mercury (II) iodide is placed in a melting point tube attached to a thermometer and heated in some form of apparatus (*e.g.*, electrical heater), it is possible to record temperature at which the red mercury (II) iodide changes to the yellow form.

Density Change

As rhombic sulphur changes to monoclinic sulphur, there is a decrease in density and, therefore, an increase in volume. The change in volume is employed to measure the transition temperature by using an apparatus known as *Dilatometer*

Some powdered rhombic sulphur is placed in the glass bulb and liquid paraffin (an inert liquid) is introduced above the sulphur. The apparatus is then immersed in a heating water-bath, the temperature of which is raised. The scale reading and the temperature is recorded every minute. A plot of liquid level in the capillary against temperature gives a curve.On cooling of the dilatometer, reverse changes take place but due to thermal lag the curve assumes. The transition temperature is taken as the mean of the respective temperatures where expansion starts (T_1) and contraction begins (T_2).

Solubility Change

Two forms of the same substance have different solubilities but at the transition point they have identical solubility. Thus, if solubility-temperature graph is plotted for the two forms, it is found to consist of two parts with a sharp break. While one part represents the solubility curve for one form, the second part represents that for the other. *At the meeting point of the two curves, the solubility of the two forms is the same and it indicates the transition temperature.* For example, in the diagram for the system sodium sulphate-water, the solubility curves of

Na_2SO_4 (rhombic) and $Na_2SO_4.10H_2O$ meet at 32.2°C. Thus, 32.2°C is the transition temperature where $Na_2SO_4.10H_2O$ changes to Na_2SO_4.

Cooling Curve Method

There is often an evolution or absorption of heat when one form passes to the other. Suppose that form A is converted into form B on heating. Now let B be allowed to cool and a curve obtained by plotting the temperature against the time. *The otherwise steady curve has a distinct break at a temperature corresponding to the transition point because here heat is evolved from B.*

Degrees of Freedom

Some properties of each phase of a system are independent of the amount of the phase present. Thus the temperature, pressure, density, molar heat capacity, and refractive index, for example, of a gas are independent of the amount of gas. Properties like these, which are characteristic of the individual phases of the system and are independent of the amounts of the phases are known as *intensive properties*. Properties like the weight and volume of a phase, which depend on the amount of the phase, are known as *extensive properties*. The latter type of property will not concern us in our study of phase equilibria.

A one-phase system of one component, for example, has a large number of intensive properties. To describe the state of such a simple system, one might measure and report values for many such properties; *i.e.,* one might report the pressure, the temperature, the density, the refractive index, the molar heat capacity, and so forth. We know *from experience*, however, that it is not necessary to specify all these properties to characterize the system completely. All the intensive properties of a sample of pure liquid water, for example, are fixed if the temperature and pressure of the sample are stated. Any two intensive properties instead of temperature and pressure might have been fixed, and one would again have found that the sample was completely characterized. Our experience tells us that only a few of the many intensive properties of a sample can be arbitrarily fixed, *i.e.,* need be specified to define the sample.

The number of *degrees of freedom* of a system is defined as *the least number of intensive variables that must be specified to fix the values of all the remaining intensive variables.*

THE TWO-COMPONENT PHASE DIAGRAMS

The simplest of the two-component phase diagrams are those for liquid systems which can break up into two liquid phases. Such systems are usually treated at some constant pressure, usually atmospheric, high enough to ensure that no vapor can occur in equilibrium with the liquid phases and over a range of temperatures high enough to ensure that no solid phases appear. If the pressure is fixed, the remaining significant variables are the temperature and composition. Diagrams are therefore made showing the phase behavior in terms of these variables. The composition is usually expressed as weight percentage of one component or as weight fraction.

Three different types of behavior are recognized. Representatives. The heavy line on each diagram bounds the region in which two liquid phases appear. That the line also gives the composition of the liquid layers can be seen by following, the addition of the second component, isobutanol, to an initial quantity of pure water at 60°C. The first additions of butanol dissolve in the water to form a single phase, and this solubility persists until the total composition of the system corresponds to point a. At this point the solubility of butanol in water is reached, and further addition produces a second layer of composition c. Thus, a total composition of b, at 60°C, results in a two phase system, the phases having compositions a and c. As the amount of the second component increases, the total composition approaches c, at which point all the water-rich layer has finally dissolved in the butanol-rich layer to give a one-phase system again.

The relative amounts of the two phases that a system with given total composition gives rise to can be calculated by the procedure. The system has a total weight w and gross weight fraction of component A designated by y. The weights of the two phases that the system breaks up into are w_1 and w_2, and these phases have weight fractions of component A of y_1 and y_2, respectively.

The total weight conservation requires

$$w = w_1 + w_2$$

and the conservation of component A requires

$$yw = y_1w_1 + y_2w_2$$

Substitution of the expression for

$$y(w_1 + w_2) = y_1w_1 + {}_2w_2$$

which rearranges to

$$\frac{w_1}{w_2} = \frac{y_2 - y}{y - y_1}$$

THREE COMPONENT SYSTEMS

A three-component system can be studied with the help of a triangular graph.

The relative amount of the three components usually presented as percentages by weight, can be shown on a triangulal plot. The corners of the triangle labeled A, B, and C correspond to the pure components A, B, and C, respectively. The side of the triangle opposite the corner labeled A, for example, implies the absence of A. Thus, the horizontal lines across the triangle show increasing percentages of A from zero at the base to 100 percent at the apex. In a similar way the percentages of B and C are given by the distances from the other two sides to the remaining two apices. From the three composition scales of the diagram the composition corresponding to any point can be read off. This procedure for handling the composition of three-component systems is possible, and the total composition is always 100 per cent, because of the geometric result that the sum of the three perpendicular distances from any point to the three sides of the triangle is equal to the height of the triangle.

As with two-component systems, the simplest three-component systems are those in which a liquid system breaks down into two phases. Over a certain range of temperature, the system acetic acid-chloroform-water is such a system. A two-phase region occurs in systems with relatively low amounts of acetic acid. A necessary part of the diagram is the tie lines through the two-phase region joining the compositions of the two phases that are in equilibrium.

Thus a total composition corresponding to point a in the two-phase region gives two phases, one of composition b and the other of composition c. A unique point on the two-phase boundary is indicated by d. This point, called the *isothermal critical point* or the *plait point*, is similar to the previously encountered critical-solution temperatures, or consolute points, in that the compositions of the two phases in equilibrium become equal at this point.

Application of the phase rule to a system corresponding to a point in the two-phase region gives

$$\Phi = C - P + 2 = 3 - 2 + 2 = 3$$

The 3 degrees of freedom can be accounted for by the pressure, the temperature, and one composition variable. Thus the composition of both phases cannot be arbitrarily fixed. If one is fixed, the tie line from that composition fixes the composition of the second phase.

Three-component systems involving solids and liquids can be introduced by considering systems of two salts and water. The simplest behavior is that where the two salts are somewhat soluble and the diagram gives the curves for the saturated-solution compositions. Such diagrams are perhaps more easily understood if tie lines are also drawn to show that the saturated solutions along DF and F are in equilibrium with the solid salts B and C, respectively. Point F corresponds to a system in which the solution is in equilibrium with both salts. Removal of water from point F moves the total composition toward the base of the triangle. The effect is to form more solid salts, which remain in equilibrium with the decreasing amount but constant concentration of saturated solution.

ZINC-CADMIUM SYSTEM

The zinc-cadmium system is an example of a metal system with a eutectic. The system can be explained as follows.

Curve AO; the Freezing Point Curve of Zinc : A represents me freezing point (or melting point) of zinc (419°C). The curve AO shows that the melting point of zinc is lowered by the addition of cadmium. The phases in equilibrium along AO are solid zinc and liquid solution of zinc and cadmium. Applying the reduced phase rule equation.

$$F' = C - P + 1 = 2 - 2 + 1$$

Thus the system solid zinc/solution is *monovariant.*

Curve BO; the Freezing Point Curve of Cadmium : The point B represents the melting point of cadmium (321°C). The curve BO shows the fall of melting point of cadmium on the addition of zinc. Along this curve, solid cadmium is in equilibrium with the liquid solution of zinc and cadmium.

Applying the reduced phase rule equation to the equilibrium cadmium/ solution,

$$F' = C - P + 1 = 2 - 2 + 1 = 1$$

Thus the degree of freedom is one and the equilibrium is monovariant.

The Eutectic Point O : The curves AO and BO meet at O. Here, solid zinc, solid cadmium and solution are in equilibrium. The number of phases is, therefore, three.

$$F = C - P + 1 = 2 - 3 + 1 = 0.$$

Thus, the system at point 0 is *nonvariant.* The point O is called the eutectic point. Both the temperature and composition being fixed, the system has no degree of freedom. The eutectic temperature is 270° and the eutectic composition is 83% cadmium 17% zinc.

The Area Above AO : In this area, both the components zinc and cadmium are present as liquid solution of varying composition. The solution of the two metals being homogeneous, constitutes one phase only. Thus the system represented by the area above AO is *bivariant.*

$$F = C - P + 1 = 2 - 1 + 1 = 2.$$

The Area Above BO. In this area, zinc and cadmium exist as liquid solution. The composition of the solution is indicated on the composition axis. The liquid solution, regardless of its composition, represents one phase only. Thus the system in the area above BO is bivariant.

$$F = C - P + 1 = 2 - 1 + 1 = 2.$$

The Effect of Cooling : If the solution in the area above AO is cooled, zinc separates as the curve itself is reached. This continues till the point 0 when the solid eutectic mixture (83% Cd + 17% Zn) separates. Similarly, the solution above BO on cooling allows the separation of cadmium. This continues till the point 0 is reached at 270°C and eutectic mixture separates. As clear from the diagram, the area below the curve AO represents zinc and solution; the area below BO shows cadmium and solution. The area below the eutectic temperature represents the system solid zinc/solid cadmium. When the solid compound AB and the liquid phase have identical composition at the maximum point on the freezing point curve, the corresponding temperature is said to be the Congruent Melting Point of the compound.

THE MAGNESIUM-ZINC SYSTEM

It is a typical 2-component system which involves the formation of an intermetallic compound $MgZn_2$. It has four phases: solid magnesium (Mg), solid zinc (Zn), solid $MgZn_2$, and the liquid solution of Mg and Nz.

The complete phase diagram of the system magnesium-zinc. It appears to be made of two simple eutectic diagrams. The one toward the left represents the eutectic system Mg-$MgZn_2$, while the one to the right the system Zn-$MgZn_2$.

The curves AC, CDE, and BE. AC is the freezing points curve of magnesium; BE is the freezing point curve of zinc; and CDE is that of the compound $MgZn_2$. The curve AC shows that the melting point of magnesium (651°C) is lowered on the addition of zinc. This continues until the point C is reached. Here a new phase, solid $MgZn_2$ appears.

The curve CD shows the increase of concentration of zinc in the melt with the rise of temperature. At the maximum point D, the composition of the melt and the solid compound becomes the same *i.e.*, $MgZn_2$. The point D, therefore, represents the melting point of $MgZn_2$ (575°C). The curve DE now shows the lowering of the melting point with the addition of zinc until the lowest point is attained. Here solid zinc appears.

The curve BE exhibits that the melting point of zinc (420°C) falls with the addition of magnesium until the point E is reached. Along the freezing point curves AC, CDE and BE, there are two phases in equilibrium *viz.*, one solid phase (Mg, Zn, or $MgZn_2$) and the other liquid phase. Applying the reduced phase rule equation, we have

$$F = C - P + 1 = 2 = 2 + 1 = 1$$

This shows that the systems Mg/liquid, Zn/liquid, and $MgZn_2$/liquid are all monovariant.

Eutectic Points C and E : There are two eutectic points in the phase diagram. The systems at the points C and E have two components and three phases in equilibrium.

	PHASES PRESENT
C	Solid Mg, solid $MgZn_2$, Liquid
E	Solid Zn, solid $MgZn_2$, Liquid.

These systems are, therefore, *nonvariant.*

$$F = C - P + 1 = 2 - 3 + 1 = 0$$

Congruent Melting Point : As already stated, the composition of the compound $MgZn_2$ and the melt at D is identical. The corresponding temperature is the congruent melting point of the compound. Here the system has two phases viz., the solid compound and the melt. Both these can be represented by one component ($MgZn_2$). Therefore the system at D is *nonvariant*,

$$F = C - P + 1 = 1 - 2 + 1 = 0.$$

The Areas : The area above the curves AC, CDE, and BE represents the solution of magnesium and zinc (the melt). The single phase system at any point in this area is *bivariant*.

The phases present in the other regions of the phase diagram are as labelled. The general form of the phase diagram of such a 2-component condensed system. Here the two components A and B are completely miscible in the liquid state, and these solutions on cooling yield only pure A or pure B as solid phases.

The diagram consists of: *Curve AC; the Freezing Point Curve of A* : The point A represents the freezing point of A. The curve AC shows that the freezing point of A falls by the edition of B to A. Thus along this curve, the solid A is in equilibrium with the liquid solution of B in A. Curve BC; the Freezing point curve of B. The point B shows the freezing point of B. The curve BC exhibits the fall of freezing point by the addition of A to B. Along this curve, the solid B is in equilibrium with the liquid solution of A in B. Applying the reduced phase rule equation to the equilibria represented by the curve AC and CB *i.e.*, solid A/solution and solid B/solution respectively, we have

$$F = C = P + 1 = 2 - 2 + 1 = 1$$

The degree of freedom is one *i.e.*, both equilibria are *monovariant.*

ADSORPTION

The phenomenon of concentration of molecules of a gas or liquid at a solid surface is called adsorption. The substance that concentrates at the surface is called Adsorbate and the solid on whose surface the concentration occurs is called the *Adsorbent.*

Often we can ignore the special properties of the materials at the surface of any solids or liquids that exist, even though we recognize that in these interphase regions materials must have properties different from those of the bulk solids or liquids.

In most chemical systems the fraction of the molecules on a surface and the free-energy difference between the surface and the bulk material are relatively small. Now systems are considered in which the surface effects are dominant. Attention is devoted principally to systems in which the molecules of a gas are concentrated on the surface of a solid. The molecules are said to be *adsorbed* on the solid surface, and this process is distinguished from the penetration of one component throughout the body of a second, called absorption. We begin with a direct study of the adsorption process and the adsorbed layer.

The goal of modern physiochemical studies of surface phenomena is the understanding of these phenomena by means of a molecular model. However, systems which have a very thin, often monomolecular layer of gas adsorbed on a complex solid adsorbent resist many of the methods developed to investigate atomic and molecular properties. However, the photoelectron (ESCA) method can be applied to reveal some of the bonding properties of adsorbed species. Infrared absorption spectra, which reveal the characteristic vibrations of the adsorbed species, can also be obtained by using infrared-transparent materials and a technique that involves total internal reflection. The current theories of surface reactions, however, are still often loosely supported. But some of the background information and treatments on which present ideas are based will be introduced.

(1) *Adsorption of a Dye by Charcoal :* If finely divided charcoal is stirred into a dilute solution of methylene blue (an organic dye), the depth of colour of the solution decreases appreciably. The dye molecules have been adsorbed by the charcoal particles.

(2) *Adsorption of a Gas by Charcoal* : If a gas (SO_2, Cl_2, NH_3) is treated with powdered charcoal in a closed vessel, the gas pressure is found to decrease. The gas molecules concentrate on charcoal surface and are said to be adsorbed.

It is convenient in the study of adsorption to recognize that absorptions can be placed in one of two categories. These categories are suggested by the possibilities of having essentially physical forces holding the gas molecules to the solid or of having chemical bonds serve that function. The categories of *physical adsorption* and *chemical adsorption*, or more commonly, *chemisorption*, thus arise. The observed characteristics of any adsorption process usually allow it to be placed in one category or the other. The different behaviours of the two types of adsorption should be recognized as having the features of a physical process, such as condensation, or those of a chemical reaction. It should be pointed out that whether more than a monolayer is being formed is not directly observable but, as we shall see, can often be deduced from experimental data.

Most interest in adsorption, and in the closely related field of heterogeneous catalysis, is in chemisorption.

Chemical adsorption is called as chemisorption. Now let us study about the difference between physical adsorption and chemisorption.

Physical Adsorption	**Chemisorption**
1. Caused by intermolecular van der Waals' forces.	1. Caused by chemical bond formation.
2. Depends on nature of gas. Easily liquefiable gases are adsorbed readily.	2. Much more specific than physical adsorption.
3. Heat of adsorption is small (about 5 kcal mol^{-1})	3. Heat of adsorption is large (20-100 kcal mol^{-1}).
4. Reversible.	4. Irreversible.
5. Occurs rapidly at low temperature; decreases with increasing temperature.	5. Increases with increase of temperature.
6. Increase of pressure increases adsorption; decrease of pressure causes desorption.	6. Change of pressure has no such effects.
7. Forms multimolecular layers on adsorbent surface.	7. Forms unimolecular layer.

Heats of Adsorption

Heat of adsorption is defined as the energy liberated when 1 gm mole of a gas is adsorbed on the solid surface. In physical adsorption, gas molecules concentrate on the solid surface. Thus, it is similar to the condensation of a gas to liquid. Therefore, adsorption like condensation is an *exothermic process*. Since the attractions between gas molecules and solid surface are due to relatively weak van der Waals' forces, heats of adsorption are small (about 5 kcal mol^{-1}).

In chemisorption the attractive forces are due to the formation of true chemical bonds. Therefore, the heats of adsorptions are large (20 to 100 kcal mol^{-1}).

Reversible Character

Physical adsorption is a reversible process. The gas adsorbed onto a solid can be removed (desorbed) under reverse conditions of temperature and pressure. Thus,

$$\text{Gas} \rightleftharpoons \text{Gas/Solid} + \text{Heat.}$$

Chemisorption, on the contrary, is not reversible because a surface compound is formed.

Effect of Temperature

Physical adsorption occurs rapidly at low temperature and decreases with increasing temperature (Le Chatelier's Principle). Chemisorption, like most chemical changes, generally increases with temperature. Thus a rise of temperature can often cause physical adsorption to change to chemisorption. Nitrogen, for example, is physically adsorbed on iron at 190°C but chemisorbed to form a nitride at 500°C.

Effect of Pressure

Since a dynamic equilibrium exists between the adsorbed gas and the gas in contact with the solid as stated in (4), Le Chatelier's Principle is applied. Actually it has been found that increase of pressure leads to increase of adsorption and decrease of pressure causes desorption.

Thickness of Adsorbed Layer of Gas

From a study of the isotherms relating to the amount of gas adsorbed to the equilibrium pressure, Langmuir showed that at low pressure, the physically adsorbed gas forms only one molecular thick layer. However, above a certain pressure, multimolecular thick layer is formed.

ADSORPTION ISOTHERM

The most common adsorption experiment is the measurement, for a given amount of absorbent, of the relation between the amount of gas adsorbed and the pressure of the gas. Such measurements are usually made at a constant temperature, and the results are generally presented graphically as an *adsorption isotherm*. Experimentally, one measures either the volume of gas taken up by a given amount of adsorbent or the change in weight of the adsorbent when it is exposed to a gas at a given pressure. The apparatus that can be used. A great variety of adsorption-isotherm shapes are found. Chemisorption is usually accompanied by an initial steeply rising curve that gradually flattens. The initial rise is taken as corresponding to the strong tendency of the surface to bind the gas molecules, and the levelling off can be attributed to the saturation of these forces, perhaps by one or more of the three mechanisms mentioned in the preceding section. Physical adsorption, however, is accompanied by an adsorption isotherm that tends to have an increasingly positive slope with increasing gas pressure. Each incremental increase in gas pressure produces a larger increase in the amount of gas adsorbed—up to the limit of a pressure equal to the vapor pressure of the material being adsorbed, at which pressure the adsorption isotherm ascends vertically as condensation occurs.

Some adsorption isotherms, suggests, can be interpreted as a combination of these Chemisorption and physical-adsorption curves.

However no simple or even complex explanation can be expected for the detailed shapes of all adsorption isotherms.

LANGMUIR ADSORPTION ISOTHERM

Langmuir (1916) derived a simple adsorption isotherm based on theoretical considerations. It was named after him.

Assumptions

Langmuir made the following assumptions.

(1) The layer of the gas adsorbed on the solid adsorbent is one-molecule thick.

(2) The adsorbed layer is uniform all over the adsorbent.

(3) There is no interaction between the adjacent adsorbed molecules.

Langmuir considered that the gas molecules strike a solid surface and are thus adsorbed. Some of these molecules then evaporate or are 'desorbed' fairly rapidly. A dynamic equilibrium is eventually established between the two opposing processes, adsorption and desorption.

If θ is the fraction of the total surface covered by the adsorbed molecules, the fraction of the naked area is $(1-\theta)$. The rate of desorption (R_d) is proportional to the covered surface 9. Therefore,

$$R_d = k_d\,\theta,$$

where k_d is the rate constant for the desorption process.

The Langmuir theory suggests that the rate of evaporation can be taken to be proportional to the fraction of the surface covered and can therefore be written as $k_1\theta$, where k_1 is some proportionality constant. This simple proportionality assumption ignores the fact that the enthalpy of adsorption generally depends on the extent of coverage. (Then the simple assumption of an evaporation rate proportional to $k_1\theta$ might not hold.)

The rate of condensation, furthermore, is taken to be proportional both to the gas pressure P, which according to the kinetic-molecular theory determines the number of molecular collisions per unit area per unit time, and to the fraction of the surface not already covered by adsorbed molecules, *i.e.*, to $1-\theta$. It is assumed that only collisions with this exposed surface can lead to the sticking of a molecule to the surface. The relation between equilibrium surface coverage and gas pressure is then obtained by equating the expressions deduced for the rate of evaporation and the rate of condensation.

This, $$k_1\theta = k_2P\,(1-\theta) \quad \text{...(1)}$$

where k_2 is another proportionality constant. Rearrangement gives

$$\theta = \frac{k_2P}{k_1+k_2P} \quad \text{...(2)}$$

Introduction of $a = \dfrac{k_1}{k_2}$ allows this result to be written as

$$\theta = \frac{P}{a+P} \quad \text{...(3)}$$

Inspection of chemisorption-type isotherm is obtained from this theory. At small values of P, where P in the denominator can be neglected compared with a, reduces to a simple proportionality between θ and P, and this behavior is that corresponding to the initial steep rise of the isotherm curve. At higher pressures the value of P in the denominator contributes appreciably, and the increasing denominator leads to values of θ that do not increase proportionally to the increase in P. For sufficiently large values of P, then, θ approaches the constant value of unity.

Experimental isotherm data consist of the amount of gas adsorbed by a given amount of adsorbent as a function of the gas pressure. For adsorption, up to a monolayer, the amount of gas y adsorbed at some pressure P and the amount of gas y_m needed to form a monolayer are related to θ according to

$$\frac{y}{y_m} = \theta \qquad ...(4)$$

Equation (3) becomes

$$y = \frac{y_m P}{a + P} \qquad ...(5)$$

We can write the equation as,

$$\frac{P}{y} = \frac{a}{y_m} + \frac{P}{y_m}$$

ADSORPTION ISOTHERM LIQUIDS

The relation between the amount of solute adsorbed from a liquid solution and the concentration is another type of adsorption isotherm.

A very important but little understood process is the adsorption of a solute of a solution onto a solid adsorbent. This procedure is followed, for example, in the decolorizing of solutions using activated charcoal. The separation technique of chromatography makes use of the relative adsorption tendencies of the solutes of a solution.

This process of adsorption from solution is even more difficult to treat theoretically than the corresponding gas-on-solid process. It appears, however, that only a monomolecular layer is formed, any further addition being strongly opposed by the solvating power of the solvent.

A fairly satisfactory empirical isotherm, which can be applied to adsorptions of gases with considerable success but has been used principally for adsorption from solution, has been given by M. Freundlich.

If y is the weight of solute adsorbed per gram of adsorbent and c is the concentration of the solute in the solution, this empirical relation is where k and n are empirical constants. The equation is conveniently used in the logarithmic form

$$\log y = \log k + 1/n \log c$$

When this equation is applied to gases, y is the amount of gas adsorbed and c is replaced by the pressure of the gas. Experimental results conform to the Freundlich expression if a plot of log y versus log c, or log P, yields a straight line. The constants can then be determined from the slope and intercept.

APPLICATIONS OF ADSORPTIONS

The phenomenon of adsorption is used in many fields and industries. Now let us study some of the applications of adsorption.

Adsorption finds application in the production of High vacua.

Gas Masks

All gas masks are devices containing an adsorbent (activated charcoal) or a series of adsorbents. These adsorbents remove poisonous gases by adsorption and thus purify the air for breathing.

Heterogeneous Catalysis

In heterogeneous catalysis, the molecules of reactants are adsorbed at the catalyst surface where they form an 'adsorption complex'. This decomposes to form the product molecules which then take off from the surface.

Removal of Colouring Matter from Solutions

Animal charcoal removes colours of solutions by adsorbing coloured impurities. Thus, in the manufacture of cane-sugar, the coloured solution is clarified by treating with animal charcoal or activated charcoal.

Flotation Process

The low grade sulphide ores (PbS, ZnS, Cu_2S) are freed from silica and other earthy matter by *Froth Flotation Process*. The finely divided ore is mixed with oil (pine oil) and agitated with water containing a detergent (foaming agent). When air is bubbled into this mixture, the air bubbles are stabilized by the detergent. These adsorb mineral particles are wetted with oil and rise to the surface. The earthy matter wetted by water settles down at the bottom.

Chromatographic Analysis

Mixtures of small quantities of organic substances can be separated with the help of *chromatography* which involves the principle of selective adsorption. The mixture is dissolved in a suitable solvent (*hexane*) and poured through a tube containing the adsorbent (alumina). The component most readily adsorbed is removed in the upper part of the tube. The next most readily adsorbed component is removed next, and so on.

Thus the material is separated into 'bands' in different parts of the tube. Now pure solvent is poured through the tube. Each component dissolved in the solvent comes down by turn and is collected in a separate receiver.